Crop Improvement for Salinity Tolerance

Edited by
Luxiang Liu

Institute of Crop Sciences,
Chinese Academy of Agricultural Sciences,
Beijing 100081,
China

China Agricultural Science and Technology Press

图书在版编目（CIP）数据

作物耐盐性改良 = Crop Improvement for Salinity Tolerance / 刘录祥等编著. --北京：中国农业科学技术出版社，2022. 11
ISBN 978-7-5116-6009-1

Ⅰ. ①作…　Ⅱ. ①刘…　Ⅲ. ①耐盐性－作物育种－良种繁育　Ⅳ. ①S33

中国版本图书馆CIP数据核字（2022）第 210796 号

责任编辑 姚　欢　施睿佳
责任校对 李向荣
责任印制 姜义伟　王思文

出 版 者 中国农业科学技术出版社
北京市中关村南大街 12 号　　邮编：100081
电　　话 （010）82106631（编辑室）　（010）82109702（发行部）
（010）82109709（读者服务部）
网　　址 https: // castp.caas.cn
经 销 者 各地新华书店
印 刷 者 北京建宏印刷有限公司
开　　本 185 mm × 260 mm　1/16
印　　张 16.25　彩插 3 面
字　　数 600 千字
版　　次 2022 年 11 月第 1 版　　2022 年 11 月第 1 次印刷
定　　价 108.00 元

INFORMATION OF THE AUTHORS

Zahoor Aslam
National Bio-Saline Agriculture Program
Ministry of Food, Agriculture and Livestock
Islamabad, Pakistan

Guilan Cao
Institute of Crop Sciences
Chinese Academy of Agricultural Sciences
Beijing 100081, China

Ruzhen Chang
Institute of Crop Sciences
Chinese Academy of Agricultural Sciences
Beijing 100081, China

Zhonghua Chen
Institute of Biomedical and Life Sciences
University of Glasgow
Glasgow, G12 8QQ, UK

Rongchao Ge
College of Life Sciences
Hebei Normal University
Shijiazhuang 050016, China

Jiayu Gu
Institute of Crop Sciences
Chinese Academy of Agricultural Sciences
Beijing 100081, China

Rongxia Guan
Institute of Crop Sciences
Chinese Academy of Agricultural Sciences
Beijing 100081, China

Guizhen Guo
Institute of Rice Research
Jilin Academy of Agricultural Sciences
Gongzhuling 136100, China

Huijun Guo
Institute of Crop Sciences
Chinese Academy of Agricultural Sciences
Beijing 100081, China

Longzhi Han
Institute of Crop Sciences
Chinese Academy of Agricultural Sciences
Beijing 100081, China

Philip Hollington
School of Natural Sciences
Bangor University
Bangor LL57 2UW, Wales, UK

Jianhua Huang
Biotechnology Research Institute
Shanghai Academy of Agricultural Sciences
Shanghai 201106, China

Zhanjing Huang
College of Life Sciences
Hebei Normal University
Shijiazhuang 050016, China

Ahsan Irshad
Institute of Crop Sciences
Chinese Academy of Agricultural Sciences
Beijing 100081, China

Si-Yong Kang
Advanced Radiation Technology Institute
Korea Atomic Energy Research Institute
Jeongeup, Korea

Don Sub Kim
Advanced Radiation Technology Institute
Korea Atomic Energy Research Institute
Jeongeup, Korea

Jin-Baek Kim
Advanced Radiation Technology Institute
Korea Atomic Energy Research Institute
Jeongeup, Korea

Geung-Joo Lee
Advanced Radiation Technology Institute
Korea Atomic Energy Research Institute
Jeongeup, Korea

Myung-chul Lee
National Institute of Agricultural Biotechnology
RDA, Suwon 441-707, Korea

Chengdao Li
Centre for Crop and Food Innovation
Food Futures Institute
Murdoch University
90 South Street, Murdoch Western Australia 6150

Luxiang Liu
Institute of Crop Sciences
Chinese Academy of Agricultural Sciences
Beijing 100081, China

Dongling Qi
Institute of Crop Sciences
Chinese Academy of Agricultural Sciences
Beijing 100081, China

Lijuan Qiu
Institute of Crop Sciences
Chinese Academy of Agricultural Sciences
Beijing 100081, China

Yinzhu Shen
College of Life Sciences
Hebei Normal University
Shijiazhuang 050016, China

Jae Young Song
Advanced Radiation Technology Institute
Korea Atomic Energy Research Institute
Jeongeup, Korea

Seok-cheol Suh
National Institute of Agricultural Biotechnology
RDA, Suwon 441-707, Korea

Lei Tian
Institute of Crop Sciences
Chinese Academy of Agricultural Sciences
Beijing 100081, China

Guoping Zhang
College of Agriculture and Biotechnology
Zhejiang University, Huajiachi Campus
Hangzhou 310029, China

Junguo Zhang
Institute of Rice Research
Jilin Academy of Agricultural Sciences
Gongzhuling 136100, China

Sanyuan Zhang
Institute of Rice Research
Jilin Academy of Agricultural Sciences
Gongzhuling 136100, China

Xikui Zhang
Institute of Crop Sciences
Chinese Academy of Agricultural Sciences
Beijing 100081, China

Baocun Zhao
College of Life Sciences
Hebei Normal University
Shijiazhuang 050016, China

Linshu Zhao
Institute of Crop Sciences
Chinese Academy of Agricultural Sciences
Beijing 100081, China

Shirong Zhao
Institute of Crop Sciences
Chinese Academy of Agricultural Sciences
Beijing 100081, China

Meixue Zhou
Tasmanian Institute of Agricultural Research
University of Tasmania
Tas 7250, Australia

Qingyang Zhou
Institute of Rice Research
Jilin Academy of Agricultural Sciences
Gongzhuling 136100, China

FOREWORD

According to the prediction of Food and Agriculture Organization of the United Nations (FAO), grain production has to be increased by more than 70% by 2050 to meet the demand. At the same time, with the elevation of consumption levels, the adjustment of dietary structure will continue to push up the demand for food globally, and food shortages in many countries will become increasingly serious. Salinity, drought and water shortages have become serious bottlenecks to increasing crop production. Soil salinization, especially in highly valuable irrigated lands is a major worldwide land degradation problem that has brought about severe soil quality decline, crop productivity losses and overall reduced environmental sustainability. This salinity problem is developing rapidly in Asia, with one-fourth of the world's population and limited land and water resources. Enhancing sustainable intensification of agricultural production in these salt-affected areas requires careful planning to diagnose the specific problems of the salt-affected soils and select the most appropriate restorative/mitigation technologies for adoption in close interaction with the farmers. In view of the several limitations of the classical soil and water management measures, innovative cost-effective approaches focus on the identification of plant genotypes better adapted to these abiotic (salinity, drought and heat) stresses and the development of new crop cultivars.

These varieties with improved tolerance to such stresses will provide the farmers added value and a competitive market advantage, which in turn will contribute to increased farm income and improved human welfare. Breeding for salt tolerance in a sustainable manner demands a thorough understanding of the genetic and physiological mechanisms of salt tolerance for identifying salt-tolerant plant germplasm. In the past, this targeted plant breeding for salt tolerance has been hindered due to the limited knowledge of the physiological salt tolerance processes and their complex genetic nature, thus resulting in the lack of efficient selection procedures.

This book covers all aspects of crop improvement for salinity tolerance, from soil-water-plant relationships in salt-affected areas including soil and water salinity and sodicity, salinization processes, effects of salinity on both soils and plants and mechanisms of crop salt tolerance, assessment and reclamation of salt-affected soils, and field experimentation in saline areas, to the genetics and progress made on breeding of major crop plants (wheat, rice, barley and soybean) for salinity tolerance. It dedicates a final part to the main experimental protocols for breeding salt-tolerant crops (wheat, barley and rice) using in vitro mutagenesis techniques, including a brief assessment of their efficiency and future prospects.

The book is aimed at all levels of administrative and technical personnel involved in plant breeding programs to abiotic stresses such as salinity, drought and heat in Asia, including Ministries of Agriculture and Rual Affairs/Ecology and Environment, directorates of Plant Production Services, faculties of Agricultures/Agro-Ecology/Biology and Plant/Soil Sciences, and institutions involved in climate change and environmental sustainability. It is hoped that the publication will assist plant production/breeding and crop improvement scientific and technical personnel to apply these novel research and development technologies, thus contributing to progress in scientific research and sustainable intensification of agricultural production in salt-affected areas of Asia and elsewhere in the world.

This publication was put together by Prof. Luxiang Liu, Institute of Crop Sciences, Chinese Academy of Agricultural Sciences(CAAS). Grateful acknowledgment was expressed to this book's authors for their hard work.

CONTENTS

CHAPTER 1.

BASIC CONCEPTS ON SALINIZATION

Zahoor Aslam

National Bio-Saline Agriculture Program (NABSAP), Ministry of Food, Agriculture and Livestock (MINFAL), Islamabad, Pakistan

1. INTRODUCTION

Salinization has been recognized as a major soil degradation process occurring worldwide, particularly in arid and semi-arid regions under both irrigated and dry land agriculture. The soil salinization may be of primary or secondary origin. While salinization of primary origin is due to the long-term influence of natural processes leading to an accumulation of salts in the arid regions, secondary salt-affected soils are the result of the salts stored in the soil profile and/or groundwater being mobilized by extra water provided by human activities such as irrigation and/or land clearing. The causes of salinity build-up are usually associated either with irrigation mismanagement or sea water intrusion near the coastal areas (Al-Ghawas, 1997; Al-Ghanem, 1997; Al-Gabri, 1997). Although the extent of secondary salinity is smaller than primary salinity, it has huge socio-economic impacts, especially in developing countries. Salinization leads to serious soil fertility decline, crop productivity losses and crop failures, resulting in economic damages and social unrest. The salinization phenomenon is not new; in the past this process has forced people to shift to other locations, which in turn became salinized in many cases (Szabolcs, 1989).

History reveals that ancient civilizations began and developed in an environment of irrigated agriculture. It also shows that salinity has been the major and ever-present worldwide threat for the sustainability of irrigated agriculture. Downfall of several civilizations has been attributed to the final blow by waterlogging and salinity, although many (and varied) detrimental factors were also at work. One notable example is Mesopotamia, where an early great civilization developed in the valley formed by the Tigris and Euphrates rivers and agriculture was the main stay of the economy. With time, waterlogging, the precursor of salinization, appeared, and the salinity problems followed. Wheat was gradually replaced by the salt-tolerant barley crop. Eventually salt reached such concentrations that even the substitute barley crop would not grow. This degradation process took almost two millennia to devastate well-flourished agriculture (Boyden, 1987). Historical evidence from other parts of the world, such as the Indus river basin (Pakistan), the Nile valley in Egypt, China, South America and Arizona in the USA also indicate that salinity problems had a great influence on these civilizations (Casey, 1972).

For some people, salinity problems have been viewed as a cancer or a divine curse whereas for others soil salinization was a problem for which a cure may be found. In fact, for thousands of years, human beings have sought solutions that they hoped would help mitigate salinity effects. In a quest to maintain farm productivity, several on-farm and off-farm interventions and miscellaneous soil additives have been employed. For instance, different plant species and crops/crop varieties were grown. In general, the choice and use of these interventions was based on empirical principles. Some of these interventions were simply shots in the dark and were abandoned after trial and error. To avoid the agony of the long process of trial and error, scientific information on soil salinization is needed to understand the processes, establish the cause-effect relationship, and conceive/develop appropriate science-based solutions.

The search for technologies dealing with the salinity problems continues today with the only difference that human beings are now better informed. Advancement in scientific knowledge has improved the ability of human beings to design better strategies and approaches. With advances in the knowledge, concepts about salinity problems are sharpened and the diverse complex nature of the problem is better understood. This has also made it possible to develop and apply more site-specific and effective solutions. Scientists/engineers and farmers are now working on a long list of possible interventions in the context of their improved concepts or perceptions. Such basic concepts related to soil, water and plants in salt-affected soils as well as perceptions/perceived solutions, mostly in the context of irrigated agriculture/secondary salinity, are briefly described below.

2. SALT-AFFECTED SOILS

The term "salt-affected" soil is being used commonly to define a soil, which contains sufficient content of soluble salts (saline) and/or exchangeable sodium (sodic) to interfere with the growth of plants. Salt-affected soils include three main broad categories: saline soil, sodic soil and saline-sodic.

2.1 Saline soil

A soil, which contains sufficient concentration of soluble salts in the root-zone to impair the growth of crop plants, is defined as "saline". However, salt injury depends on plant species/ variety, growth stage, environmental factors, and nature/concentration of the salts. Therefore, it is very difficult to define a saline soil precisely. The most widely accepted definition of a saline soil is one that gives an electrical conductivity in the soil saturation extract (EC_e) exceeding 4 dS m^{-1} at 25° C. Saline soils have pH usually < 8.5, EC_e > 4 dS m^{-1} and exchangeable sodium percentage (ESP) < 15%. The high EC_e with low ESP tends to flocculate the fine soil particles into aggregates. The permeability is either greater or equal to those of similar normal soils.

In salt-affected areas, though the NaCl is the predominant salt, the occurrence of soluble compounds of sodium, potassium, calcium, magnesium, iron, boron, chlorides, sulfates, bicarbonates and carbonates have been reported in both soil and water. Their accumulation is usually recognized by the presence of a white salt-crust during some part of the year.

2.2 Sodic soil

Sodicity is a measure of concentration of sodium ions in the soil exchange complex. It is expressed as ESP when related to the total cation exchange capacity of the soil or as sodium adsorption ratio (SAR) relative to calcium and magnesium ions. Sodic soils contain ESP >15% and EC_e < 4 dS m^{-1} and pH generally range between 8.5 and 10 and may be even as high as 11. The low EC_e with high ESP tends to cause dispersion of the fine soil particles and destruction of soil aggregates and hence, resulting in lower soil permeability.

2.3 Saline-sodic soil

Saline-sodic soils contain both sufficient concentration of soluble salts (EC_e > 4 dS m^{-1}) to interfere with the growth of most crop plants and sufficient sodium (ESP > 15%) to affect the soil physical properties and plant growth adversely by the degradation of the soil structure. The pH of these soils may be less or more than 8.5.

Sometimes it is advantageous to further refine the categories of salt-affected soils to distinguish the intensity of the salinity problem as follows: class 0 (non-saline, 0–2 dS m^{-1}), class 1 (very slightly saline, 2–4 dS m^{-1}), class 2 (slightly saline, 4–8 dS m^{-1}), class 3 (moderately saline, 8–16 dS m^{-1}) and class 4 (strongly saline, > 16 dS m^{-1}). Class 0 shows no visible salts on the soil surface and plant growth is not affected by salinity/sodicity. In class 1 & 2 soils the plant growth may be uneven or patchy. Salts are generally presented in small sized patches, which does not cover more than 25% area collectively. In class 3 the plant growth on these soils is very patchy and the salts are fairly visible on the soil surface. The area in class 4 lies unused and may support some salt-tolerant plants.

3. EXTENT OF SALT-AFFECTED SOILS

As a general estimate salt-affected soils cover about 7% of the total soil surface of the world and no continent is free from salt-affected soils with the following approximate distribution (Szabolcs, 1989): Australia 38%, Asia (North, Central, South, Southeast) 34%, South America 14%, Africa 8.6%, North America 1.7%, Central America 0.2% and Europe 3.3%. On a global level, the total extent of salt-affected soils is 955 million hectares (Mha). Other estimates show that the extent of world's irrigated land damaged by the salinity is 25% (Postel, 1990), and up to 50% (Adams and Hughes, 1990).

Other studies have reported that about 77 Mha of cultivated lands are salt-affected by human-induced processes (Oldeman et al., 1991; Ghassemi et al., 1995) and approximately more than 30 Mha can be attributed to secondary salinization of non-irrigated lands. The secondary salinization is predominantly located in the arid and semi-arid regions including Egypt, Iran, Iraq, India, China, Chile, Argentina, Central Asia, Russia, Spain, Thailand, Pakistan, Syria, Turkey, Algeria, Tunisia, Sudan and the Gulf States.

The deleterious effect of the increased salinity in the soil is to impair the plant growth. Owing to the high salt concentration of the soil, the osmotic pressure (OP) of the soil solution increases and plants are not able to extract water as easily as they can do from a relatively non-saline soil (Al-Mehrizi, 1997). Soluble salts exert this potential over and above the matrix potential already existing in the soil. Thus, as the salt concentration increases the soil water becomes less available to the plant even though the soil may contain water and appears quite moist. There are, however, cases when some plants, particularly stone-fruits are damaged at so low salt concentrations that the osmotic effects of the salts appear irrelevant. In such cases the internal salt concentration seems to damage the plants. Hence, a high concentration of salts in the root-zone may affect plant growth by complex processes broadly termed as water deficit and/or specific ion excess (Greenway and Munns, 1980).

In sodic soils, the main damage to plants is caused by their poor physical properties. The effect of Na^+ is to disperse the fine clay particles and clog the soil pores. The soil tends to swell and its aggregates to slake down, thus a desirable crumb structure collapses creating a less permeable condition. The resulting poor soil physical conditions restrict water penetration and aeration, which badly affects the growth and development of plants (Hillel, 1990).

4. ECONOMIC, ENVIRONMENTAL AND SOCIAL IMPACTS

Soil salinity may have economic, environmental and social impacts as follows.

4.1 Economic impacts

Economic impacts may be due to the following reasons.

- saline water tables can cause productive land to become barren, resulting in loss of agricultural production
- soil salinity also enhances soil "erosion" and loss of farm income
- salinity can deteriorate the quality of drinking water
- in salt-affected areas, infrastructural works such as roads and building foundations are weakened by high salty water tables
- high water table also affects biological activity in the soil
- increased costs associated with land rehabilitation for public utilities

4.2 Environmental impacts

Environmental impacts are introduced through land degradation by physical, chemical and biological changes in the soils and waterways, loss of vegetation and changes of the landscape in catchments.

4.3 Social impacts

Social impacts include increased production costs on farms and reduced value of land as a result of soil salinization. The people from the salinized area move away and the social structures/ set up are disturbed.

5. SOIL SALINIZATION DUE TO IRRIGATION PRACTICES

Irrigation mismanagement is the major cause of secondary salinity. In arid and semi-arid regions, irrigation is a must for permanent agriculture since rainfall is low. It is irrigation water, which provides control over soil moisture to meet the water requirements of different crops. Therefore, irrigation can lead to substantial increases in land productivity. Irrigated land, although small in size, accounts for a large proportion of the crop production in the world. There are many issues associated with irrigation that, if not properly understood, planned and managed, can cause build-up of soil salinization and may create other serious hazards to the environment.

There are at least two ways in which irrigation mismanagement may lead to the salinization process. Firstly, under conditions of insufficient precipitation, salts formed *in situ* by weathering of soil minerals or salt deposited from inadequately applied water tend to accumulate in the soil profile. In some areas the general shortage of good quality irrigation water necessitates the use of underlain brackish/saline water having a variable range of salinity and sodicity levels. This intensifies the problem pertaining to salinity and sodicity, and may cause a perched layer if drainage is restricted. To prevent salts build-up, a net removal of the salts out of the root-zone is mandatory.

Secondly, when too much water is applied, in addition to the losses and inefficiency in water use, the extra water raises the water table or increases the pressure of confined aquifers, creating an upward leakage of water. When the water table becomes close to the soil surface, there is waterlogging. Water is evaporated leaving the soluble salts behind, causing soil salinization and degradation. It is, therefore, wise to state that waterlogging and associated soil salinization of agricultural lands are a symptom of irrigation misuse and mismanagement that jeopardizes the integrity of soil's self-regulatory capacity. Prolonged irrigation mismanagement practices will cause irrigated soils to be more affected by excess soluble salts. Thus, from the above when irrigation water is applied, there is a two-sided question: (i) whether sufficient supplemental water is given to provide the required leaching of salts; and (ii) whether the drainage network has the capacity to remove sufficient water with its dissolved salts.

The process of sodification may occur associated with the soil salinization. The soils become sodic when the clay fraction becomes saturated with Na^+. Therefore, one of the major concerns of irrigated agriculture is the development of both salinity and sodicity in the soil and water resources. Once waterlogging and soil salinity are developed, they undermine the natural resource base by decreasing soil and water quality.

6. SALINE SEEPAGE

In several countries, saline seeps are instrumental in the development of salinity problems. A saline seep is primarily the result of the discharge of a saline groundwater system. The process is often accelerated by dryland farming, which allows water to move through salt-laden substrata below the root-zone. Saline seepage is related to intermittent or continuous saline water discharge at/or near the soil surface down slope from recharge areas under dryland conditions (Brown et al., 1983). The growth of crops is reduced or eliminated in the discharge area due to increased soluble concentration of salt in the root-zone. The characteristics and causes of saline seeps are similar everywhere. Typically, the native vegetation, which may include grasses, has been replaced with agricultural fields and cropping systems with lower potential evapo-transpiration (ET) requirements.

The water balance of a catchment can be manipulated to some extent by altering land use and agronomic practices. Varying agronomic practices should result in different amounts of water being removed from the landscape via ET. Moreover, the recharge rate under different crops depends, among other things, on soil type, irrigation and the average annual rainfall and its seasonal distribution. There is a range of agricultural practices, which can reduce the rate of recharge to the groundwater system. Specific techniques include the replacement of shallow-rooted agricultural species with deep-rooted ones such as alfalfa and perennials (Schofield et al., 1989).

7. MANAGEMENT OF SALINITY PROBLEMS

As mentioned earlier, secondary salinity is not a recent phenomenon. It is as old as irrigation. Salinization forced early settlers in Mesopotamia, the Indus river basin and China to abandon their degraded lands and move to non-salinized land. Although many irrigation systems around the world are confronted with serious problems of secondary salinity, the abandonment option is no longer possible. The development of the world's irrigated land and water resources has demanded a huge capital investment and massive engineering efforts. The gigantic irrigation system in the Indus river basin of Pakistan with 16 Mha of irrigated land is one example of such world's largest infrastructural works. Further development of new land and water resources will be much more difficult and expensive than in the past. Therefore, an integrated approach involving the efficient use of available irrigation and land resources together with the reclamation/improvement and sustainable utilization of saline land and water resources are better long-term options to be considered (Mashali, 1995).

A wide range of management options is available for sound management of salinity in irrigated land and dryland (non-irrigated) as well as preventing salinization of watercourses. However, implementation of any option depends on highly site-specific conditions. While some of the options are more suitable for irrigated land salinity, others are suitable for dryland salinity or for reducing salinity levels in watercourses. Technical, economic, social and political considerations are major factors influencing the selection and implementation of management options.

7.1 Salinity control and methods of irrigation

In arid and semi-arid zones the major constraints to agricultural production are (i) limited quantities of fresh water resources, and the need to increase its use efficiency; and (ii) exploitation of unsuitable (low quality) saline water for irrigation. The method of irrigation affects both the efficiency of water use and the way salt accumulates. Therefore, there is a clear need to adopt suitable irrigation techniques for eliminating salinity hazards to the soil. Each irrigation method has certain advantages and disadvantages and all known factors should be considered before attempting to improve salinity control by changing the method (Ayers and Westcot, 1985).

7.1.1 Irrigation channel lining or piping

Eliminating channel seepage can have some effects on the problems of shallow water tables and salinity. Hawkins (1978) describes different types of channel lining, which include concrete lining, masonry-type lining, synthetic membrane lining, bentonite membrane lining, and compacted earth lining. Channels can also be replaced by pipelines. Piping is usually considered an economic proposition in areas with permeable soils and high seepage losses.

7.1.2 Surface irrigation

Flood, basin, border and furrow are known surface irrigation methods. At the end of each irrigation cycle, the soil dries out and the salt concentrates, which adversely affects the crop yields. With surface irrigation methods, the depth of the applied water layer entering the soil varies with the location in the field and depends on the infiltration rate and time available for infiltration. Surface irrigation methods are usually not sufficiently flexible to apply less than 80 mm layer depth per irrigation. Irrigating more frequently to reduce salinity and possible water stress may also waste water and increase waterlogging and drainage problems.

Differences in the infiltration rate may be caused by land slope, degree of compaction, textural changes and soil chemistry. The time during which infiltration can take place also varies; the upper end of the field nearest the water source usually has water on its surface for a much longer time than the lower end does. High elevation spots in the field receive less water because they are covered by less water and for a shorter period. There may also be areas of more dense soil where water does not infiltrate sufficiently to provoke leaching. In such cases isolated pockets of accumulated salts are frequently formed. Surface flooding may apply a uniform depth of water layer across the entire field, thus salt accumulation increases with depth. Moreover, in the case of furrow irrigation, which applies water to only part of the field surface, salts, may also accumulate in the ridges between furrows.

7.1.3 Pressurized irrigation systems

The alternatives to improve the efficiency of surface irrigation methods are the pressurized irrigation systems, drip or sprinkler irrigation. Such systems must meet the crop requirements for water ET thus it may be easier to increase the frequency of irrigation to relieve water stress with such systems rather than with surface flooding. They often enable efficient and economic use of water, and reduce deep percolation losses. If the water application through pressurized irrigation systems is in close agreement with crop needs (ET and leaching), drainage and high water table problems can be greatly reduced, which in turn should improve salinity control. However, pressurized irrigation systems have their problems too and are not adapted to all conditions of water, soil, climate or type of crop (Ayers and Westcot, 1985). In developing countries, the change over to pressurized irrigation systems may be too expensive and will require justification, and better crop adaptability.

7.1.3.1 Drip (Trickle) irrigation

It is a method that supplies the required quantity of water to the crop almost on a daily basis. Drip irrigation has the potential to sustainable utilize saline and/or sodic waters. The poor quality water used in drip irrigation may yield better due to continuous high moisture contents and daily replenishment of water loss by ET. Salts may accumulate with drip irrigation at the surface, within the soil, outside edges of the area wetted by emitters but the daily irrigation continuously moves the water down to keep the salts under control. Therefore, frequent applications of water in small quantities could be an effective way to manage salinity and sodicity hazard of irrigation water.

In case the water is saline, frequent applications of water would keep the salts in the root-zone of the plants close to that in the irrigation water and salt would move and accumulate at the periphery of the wet zone preferably away from the roots. In case the water is sodic, it may increase soil compaction reducing its permeability and cause temporary waterlogging when wet, which in turn would affect most crops adversely and reduce the flow of water towards the roots. Frequent applications of water in small amounts, would keep the soil around optimum moisture level (field capacity) reducing possibilities of both temporary waterlogging and soil compaction. The possibility of sodification and thus, increased soil compaction has been further overcome by acidification of sodic groundwater in our recent experiments (unpublished data). Moreover, frequent irrigations with drip irrigation could also reduce the precipitation of $CaCO_3$ by keeping the soils wet.

7.1.3.2 Sprinkler irrigation

It is a method of making rain over the canopy of plants. Sprinkler irrigation, applies a uniform depth of water layer across the entire field, thus salt accumulation increases with depth. In most cases it is not advisable to use saline water with sprinkler irrigation as it may cause leaf burn, and defoliation of salt-sensitive species.

7.1.4 Irrigation scheduling

Irrigation scheduling is the management practice of determining how much water to apply during irrigation and the timing of the application. Irrigation scheduling requires the knowledge of the crop water requirements for particular growing conditions. The amount of applied irrigation water depends on crop ET, cropping pattern and type of planting (density or spacing), leaching requirement (LR), irrigation method and effective precipitation. Correct irrigation restores the soil water deficit and avoids the application of a wasteful and potentially harmful water excess. Several methods are used for determining the time when a crop requires the application of irrigation water. Among these, the soil moisture status and the plant water deficit stress indicators are the most frequently used.

Irrigation scheduling by soil moisture measurement is probably the oldest method in existence (Campbell and Campbell, 1982). Several devices and procedures have been used for obtaining soil moisture measurements. The most frequently used are the tensiometers and the neutron probe. Tensiometers consist of a metal or plastic shaft, the lower end of which is connected to a porous ceramic cell and the upper end to a pressure-measuring device. Through the close contact of the water-saturated porous ceramic cell with the soil, the soil water pressure is transmitted via the ceramic cell from the tensiometer fluid to the pressure indicator (Albert et al., 1987). The tensiometers have the advantage that they are relatively simple and inexpensive. Their disadvantage is that they measure the soil moisture only in the immediate vicinity of the unit, so that several tensiometers are needed to give a reliable spatial average of the soil moisture in the field.

The neutron probe operates by producing fast neutrons from a radioactive source. The emitted fast neutrons are moderated (slowed down) by the water in the soil. The probe is inserted into the soil through an access tube, and the water content is directly related to the number of slow neutrons scattered back to a detector. The measurement is, therefore, an average for the soil volume surrounding the access tube, so a single measurement gives the same information that several measurements from tensiometers would give. The calibration for the probe is relatively constant from soil to soil and, once it is known, it should not change with time. Readings are rapid and the instruments are reasonably portable (Campbell and Campbell, 1982).

Irrigation scheduling does not require the knowledge of the average water content or water potential for a field. Since the field is irrigated as a unit, a single representative monitoring site can be used to indicate the water status for the entire field. It is desirable to select such a site as follows: (i) with good accessibility; (ii) located more than 20 m from any edge of the field; (iii) among healthy, vigorous plants; (iv) in soil with moisture above-average field capacity; and (v) in a location exposed to normal climatic and irrigation variations. For irrigation scheduling it is necessary to determine the full and refill points of the field. The full point is the field capacity for the soil whilst the refill point is the potential below which crop production is measurably reduced (Campbell and Campbell, 1982).

Since soil moisture measurements are generally not read on a continuous basis, moisture measurements should be plotted as a function of time so that soil moisture can be extrapolated to the point in time when irrigation is required. Irrigation should be scheduled so that the soil water content stays between the refill and full point values. If the soil water content is above the full point, leaching will occur. If it is below the refill point, crop production will be reduced (Campbell and Campbell, 1982; Boyle Engineering Corporation, 1986).

Plant water deficit stress indicators can be used for irrigation scheduling. Several observational methods can be used to asses such plant stress including changes in plant color or wilting conditions. Generally, when plant stress symptoms are observed, irrigation will be too late to prevent some reduction of plant growth and yield. Recent techniques such as the pressure chamber technique or the measurement of plant leaf/canopy temperatures with an infrared thermometer can be used to determine the plant water deficit stress (Boyle Engineering Corporation, 1986).

Leaf and canopy temperatures may be either warmer or cooler than the air, depending upon environmental factors. In humid climates canopy temperatures will be near to or higher than air temperature, with only a small range of temperatures. In arid areas, however, canopy temperatures may be more than 10℃ below air temperature and have a range of perhaps 15℃ (Jackson, 1982). It is in the arid areas where irrigation is practiced that these temperature-based techniques can work best and are most needed.

Recent advances in infrared (IR) technology have enabled the production of light weight hand-held IR thermometers that can be used to make rapid measurements of plant canopy temperatures (Jackson, 1982). The plant temperature, the air temperature and the wet and dry bulb temperatures are all used together to determine when a plant is stressed. The measurements take place inside the field where the temperatures will not be affected by the field edges, which are drier and hence warmer. Measurements are best done between noon and 2 p.m. when the sun is higher overhead, shadows are at a minimum and transpiration is at its peak.

7.2 Reclamation of salt-affected soils

In order to reclaim the existing salt-affected areas for its sustainable future use and to avoid other areas to become barren lands, there is a need to determine the best soil reclamation and management practices with particular focus on the associated problems of the salt-affected soils. Three well-known categories of reclamation methods are described below.

7.2.1 Physical methods

This category of methods includes land leveling, surface flushing, salts scraping, deep plowing and land-forming or tillage, subsoiling and sanding. With land leveling, the existing land topography is changed by mechanical shifting of soil to create a defined slope gradient. It is the process of smoothing the land to achieve a uniform surface. This can occur without significant land-forming. In irrigated areas both processes are now being aided by laser technology, which improves accuracy and precision. The overall result of land-forming and grading is that more efficient irrigation is possible. The application of water to the land surface is much more precise and the land can be drained more effectively. Land leveling is a pre-requisite for efficient leaching and/or flushing the salts from a salt-affected soil. Accessions to the groundwater are thereby reduced by the elimination of waterlogged areas, resulting in a lowering of the water table and greater control of the salinity.

Surface flushing allows the surface accumulated salts to be washed away. This technique is particularly useful where surface salt crusts are present in arid and semi-arid areas. The flushed saline water then enters the drainage system, which becomes concentrated with salts. The surface salt crusts can also be removed by mechanical means such as scraping. This practice minimizes the salts temporarily, because they can be accumulated again if there is a continuous subsurface movement of saline water to the surface.

Low salinity in the root-zone can be achieved by manipulating the soil surface conditions, i.e. bed shape/elevation and irrigation management. It is very well recognized that salts tend to accumulate on the ridges away from the wet zone when furrow irrigation is adopted. Placing the seeds on off-center slope of the single row will put the seed in minimum salinity and optimal moisture conditions. Subsoiling is particularly important to disrupt the dense soil layers at depth and enhance its permeability. This method is important while reclaiming sodic soils after the addition of suitable amendments such as gypsum and watering the field. This will enhance the removal of exchangeable sodium already exchanged by Ca^{2+} by downward movement into the lower soil layers before it is displaced into the main drainage system. Sand mixing with heavy textured soils can change the textural class permanently, and the soils become more permeable and easy to reclaim. Sanding also provides a more favorable environment for plant growth compared to the original clayey soil without sanding.

7.2.2 Chemical methods

Chemical approaches are commonly used to reclaim saline-sodic or sodic soils. To achieve adequate crop growth and production on these soils, the ESP of the soil must be lowered to safer limits, which can be achieved by adding suitable chemical amendments (Abrol et al., 1988). The main aim is to replace exchangeable sodium by calcium ions. The main amendments are the following chemicals: gypsum ($CaSO_4$), calcium chloride ($CaCl_2$), iron sulfates ($FeSO_4$), aluminium sulfates [$A1_2(SO_4)_3$], sulfuric acid (H_2SO_4), hydrochloric acid (HCl), sulphur (S) and pyrite (FeS_2).

The chemical reaction between the added amendment and the sodium in the soils leads to the formation of leachable sodium components such as sodium sulfate (Na_2SO_4) and sodium chloride (NaCl). The choice of an amendment depends on its effectiveness, relative cost and the time required for the amendment to react with the sodium in the soil. Calcium chloride, iron sulfates and aluminium sulfates are usually expensive and have not been used for the large-scale improvement of sodic soils. Amendments like sulfur and pyrite must first be oxidized to sulfuric acid by soil micro-organisms before being available for reaction.

Gypsum is the most commonly used amendment worldwide because it is the cheapest and the most abundantly available product. It comes both from mines and as a by-product of the phosphate fertilizer industry. Gypsum is soluble in water to the extent of about 0.25% at 25℃ and is a source of soluble calcium. Gypsum reacts with both sodium carbonate (Na_2CO_3) and adsorbed (exchangeable) sodium and produces sodium sulfate (Na_2SO_4), which is leachable. The gypsum requirement is the amount of gypsum required to reduce the sodium saturation to some acceptable level for a given quantity of soil. For each milliequivalent of exchangeable sodium per 100 g of soil, 0.086 g of gypsum is required. Amendments like gypsum are normally applied over the soil surface and then incorporated with the soil by disking or plowing. Gypsum mixed with the surface 15 cm of soil is more effective in the removal of exchangeable sodium than gypsum applied over the soil surface.

Mined gypsum requires grinding before it applies for the reclamation of sodic soils. However, the fineness to which gypsum must be ground is an economic consideration. Although it is often maintained that finer gypsum particles are more effective, very fine grinding entails a higher cost. Abrol et al. (1988) demonstrated that gypsum passed through a 2 mm sieve and with a wide particle size distribution is likely to be more efficient for the reclamation of sodic soils having appreciable quantities of sodium carbonate.

Whilst gypsum is used in sodic and saline-sodic soils, amendments such as sulfur, sulfuric and hydrochloric acid are only applied to calcareous sodic and/or calcareous saline-sodic soils. These chemicals lower the soil reaction (pH), react with carbonates and replace exchangeable sodium with calcium. The long-term objective is to replace exchangeable sodium with calcium, and use organic matter to bind the soil particles and improve its structure. The calcium causes particles to form clusters (flocculates), forming a very clear puddle of water.

Prior to the gypsum application to sodic soils, soluble salts, e.g. carbonates and bicarbonates should be leached out. Otherwise, gypsum equivalent to the quantities of soluble carbonates and bicarbonates will be inactivated and the reclamation process will be affected. Gypsum should be broadcasted and incorporated in the upper 10 cm soil. Water is to be ponded in the soil (up to 80% gypsum get dissolved in 7–10 days ponding water). In wheat/barley or other crops, several irrigations of 7.5 cm each are recommended to dissolve gypsum and leach salts.

The amounts (in t) of various amendments equivalent to 1 t of gypsum (in terms of 100% pure gypsum) for supplying Ca are calcium chloride ($CaCl_2 \cdot 2H_2O$) 0.85 t; calcium nitrate [$Ca(NO_3)_2 \cdot 2H_2O$] 1.06 t; press-mud (lime-sulfur 9% Ca, 4% S) 0.78 t; sulfuric acid (H_2SO_4) 0.61 t; iron sulfate ($FeSO_4 \cdot 7H_2O$) 1.62 t; ferric sulfate [$Fe_2(SO_4)_3 \cdot 9H_2O$] 1.09 t; aluminium sulfate [$Al_2(SO_4)_3 \cdot 18H_2O$] 1.29 t; elemental sulfur (S) 0.19 t; pyrites (FeS_2, 30% S) 0.63 t; and limestone ($CaCO_3$) 0.58 t. The amendments alternative to gypsum should be applied with care and based on the soil properties.

7.2.3 Biological methods

The addition of organic materials (preferably legume residues) is beneficial in terms of improving soil structure, supplying plant nutrients, preventing soil erosion and hastening reclamation. Most organic amendments lower the soil pH. Care should be taken in selecting the organic amendments, as some may cause temporary N deficiency (high C/N ratio materials), and/or increase the salinity (cow dung slurry). Growing legumes improve the soil structure and its physical properties like water holding capacity and infiltration rate are improved. Green manuring has better chances than farmyard manure (FYM) of being integrated into a reclamation management package.

Reclamation can also be triggered by growing plants on the problem soils. Along with leaf litter, they add a considerable amount of organic matter as root turnover and plant residues to the soil, and release carbon dioxide by respiration and decomposition of organic materials, which converts the insoluble calcium carbonate to bicarbonate. Calcium thus solubilized would exchange with sodium, producing sodium bicarbonate, which is soluble in water and can be easily leached down. Exchange of Na^+ with Ca^{2+} along with action by organic matter results in production of soil aggregates and restores the soil structure. This phenomenon and the mechanical action of the roots increase soil permeability and help to open up the impermeable sodic soils enhancing downward movement of water and leaching of salts.

In Pakistan the growth of kallar grass is known to improve soil conditions due to its profuse and deep rooting system to the extent that has enabled some moderately salt-tolerant crops, and a few selected varieties of barley and some other plant species to grow on the ameliorated soils.

Salt removal by agronomic crops is insufficient to maintain salt balance at intermediate levels. Under saline conditions alfalfa contains 3%–4% salts (Francois, 1981). Oster (1984) reported if 2,000 mm of irrigation water with an EC_e of 1 dS m^{-1} was applied annually to grow alfalfa, the total salt applied would be about 13 Mg ha^{-1}, and the salt uptake by alfalfa would represent 6% of the applied amount. Similar calculations made with halophytes also substantiate this point although halophytes would remove more salts than non-halophytes.

7.3 Leaching and drainage (hydro-engineering)

All irrigation projects need leaching and drainage if they are to be sustainable over time. In irrigated arid/semi-arid areas, the objective of the drainage is to avoid salinization. In humid areas it is to provide aeration for plant roots.

7.3.1 Leaching

Leaching is a traditional method for reclaiming saline soils. It is achieved by flooding or by ponding water and to avoid re-salinization, a downward flux of water is maintained through and beyond the root-zone. When the amount of applied water is greater than that used by a crop during the growing season and adequate drainage exists, the net water movement is downward and the salts are leached from the root-zone. The salt concentration in the root-zone will increase if the net downward movement of salts is lesser than the salt input from irrigation. Therefore, the salt balance must be kept under control/maintained during the cropping season in order to avoid salinity build-up. Crop water requirements and irrigation water quality are two primary parameters that have to be considered in order to ensure proper water management for salinity control. Over time, the amount of salt removed by leaching must suffice to prevent the build-up of salinity beyond the level the crop can tolerate. Therefore, it is very important to know the LR for a particular soil in order to effectively leach the salts.

The LR is the calculated fraction (depth) or quantity of water that must pass through the root-zone to maintain the EC_e of the drainage water at or below some specified level. The calculation of the LR requires some understanding of how plant growth responds to a variable salinity with depth. In general, the depth of soil leached is roughly equal to the depth of water infiltrated during leaching. The recent trend is to minimize the LR in order to prevent the rising of groundwater and minimize the load in the drainage system. The LR to avoid salt damage to crop yields can be obtained and also the threshold salinity level of the crops (Maas, 1990). Estimates of the average LR for a field based on salinity measurements would require repeated measurements at several locations (Rhoades, 1980). As a general rule, the more salt-tolerant is the plant, the lower would be the LR (Aslam, 1985).

Timing of leaching does not appear to be critical, provided the crop salinity tolerance limit is not exceeded for extended periods of time or the leaching is done during critical plant growth stages. The leaching can be achieved at each irrigation time. However, in a soil with low permeability and for crops sensitive to excess moisture in the root-zone, leaching at each irrigation time may not be possible. Leaching should be done when the soil moisture is low and the water table is deep and it should precede the critical growth stage of the crop. Moreover, leaching is recommended when the demand for ET is low, i.e. at night, during high humidity, or in cooler weather. Leaching should be preferably done before or at the end of the cropping season.

The salinity data obtained during several field experiments (US Salinity Laboratory Staff, 1981) indicate that salinity changes during a crop season do provide useful information for water management. Where the water and salts have moved upward into a soil profile over a period of time, one would expect to find an "inverted" salinity profile or a soil profile in which the salinity decreases with depth; and where the dominant movement of water is downward one normally would encounter a "normal" salinity profile (Shahid and Jenkins, 1992). It is emphasized that an excellent on-farm leaching/drainage system has no value unless there is an adequate outlet for the disposal of the drainage water.

7.3.2 Drainage for salinity control

Drainage implies the removal of salts in the drainage water. Presumably drainage is required to remove water applied in excess of crop use, so that it does not transport salts to soil surface by capillary action. Not all the water requiring removal by drainage may originate from on-farm irrigation. In fact, it is common that a significant part of the water input to the system is the result of seepage from irrigation canals or from areas irrigated some distance above the area under consideration.

The traditional drainage system lowers the water table, minimizes upward water flow and thus minimizes the build-up of salts due to capillary rise, and also provides adequate leaching. Under conditions where the underlying layers are permeable and the relief is adequate, natural drainage may work. However, these ideal conditions generally do not exist in saline areas and therefore, a drainage system is always required. Depending upon the site condition and nature of the problem different types of drainage systems can be used as follows.

1) Surface drainage

In this technique the excess water is removed by shaping the land so as to make the water flow over the surface to furrows, ditches or waterways. Excess runoff water on farmlands may be caused by over-irrigation, high rainfall or poor land grading. Surface drainage systems are used to overcome these conditions. Surface drains may be deep enough to intercept groundwater, which will then enter the drain. The rate of groundwater flow into the drain depends on soil permeability and the height of the water table. The deeper a drain, the greater the width of adjacent land affected by the drawdown of the groundwater. Although surface drains can serve to control the water table, their use is being superseded by subsurface drains. Disadvantages of surface drains include loss of land, hindrance to farming operations and overland traffic, and heavy maintenance requirements due to the prolific weed growth or instability of the banks (Zijlstra and van Someren, 1980).

2) Subsurface drainage

This technique controls the water table at a safer depth, consisting of open ditches or tile drains or perforated plastic pipes, mole drainage, and vertical drainage (water pumping) when the deep horizons have an adequate hydraulic conductivity.

The seepage interceptor drain is a suitable intervention when a seep or spring is fed by a distinct water source. Interceptor drains would cut off this excess groundwater before it reaches the problem area. Interceptor drains are usually single tube or open drains placed between the water source and the problem area. Accurate location of the source of excess groundwater and proper placement of the drain are critical to the success of this form of drainage. However, investigation of subsoil conditions and groundwater flow is essential before drainage work begins (George, 1985).

In the horizontal subsurface drainage the water table and salinization are controlled by the installation of horizontal drains at a certain depth (1.5–2.5 m) below the surface. The pattern of drains allows the wet soil to drain to collectors, which remove the excess water from the land. The water flow into drains is induced by the lateral hydraulic gradient. The drains are installed at an appropriate depth and are spaced at intervals designed to ensure that the water table in the intervening space does not rise above a given height. Please see McWhorter (1977), FAO (1980), Euroconsult (1989) and Amer (1990) for more information on the design of subsurface drainage systems.

A mole drain is a simple subsurface drain. For its construction, a small mole (25–100 mm) is drawn through the ground at a pre-determined level (0.4–0.7 m and spacing of 2–5 m). In the right soil, this mole then creates a cavity similar to the inside of a pipe that will allow the passage of water. Since no pipe or refill material is required, this is the lowest priced of all drainage techniques; likewise, it is expected to have the shortest length life and possibly the lowest efficiency (Hawkins, 1978). The practical application of a mole drain is limited to heavy clay soils. Mole drains have limited stability and the moling operation has to be repeated every few years (Castle et al., 1984).

Vertical drainage systems are basically water wells spatially located on a grid which, like horizontal drainage, bring the water table down to a pre-determined level. Groundwater pumping is the most effective method of lowering the water table and removing groundwater. Tube-well drainage may be an alternative to horizontal drainage in areas with productive aquifers at moderate depth. Tube-well drainage offers the following advantages (Euroconsult, 1989): (i) it can be applied on undulating land without extensive earth movement and leveling for the installation of pipelines or main drainage channels to interconnect the wells; (ii) it diminishes maintenance costs because of the smaller network of canals and/or drains that are necessary; (iii) the water table can be drawn down to a much greater depth, reducing the risk of salinization of the soil; and (iv) it diminishes artesian pressure of aquifers underlying the top layer, which may reverse the direction of flow so that downward percolation occurs. In areas where groundwater is saline, the effluent of drainage wells is likely to be much more saline than that of horizontal drains, which makes vertical drainage less attractive or not acceptable for those areas.

Hydro-geological investigations are necessary for the feasibility study and design of pumping wells or well-point systems. The main objectives are the determination of the extent, depth and magnitude of the aquifer, its transmissivity and the vertical hydraulic conductivity of the covering layers and the effect of pumping from deeper aquifers on shallow water tables (Driscoll, 1989). To be able to pump groundwater, there is a need to have a pocket of very coarse sand or gravel (an aquifer) below the surface into which a slotted pipe ("well-point" or "spear") can be installed. The water drains into pipes through the slots and can then be pumped to the surface. These systems are also sometimes referred to as "bores". Apart from the salinity and water table control, pumping groundwater can provide extra water to supplement irrigation supplies (depending on the salinity of the groundwater). Pumping groundwater from shallow aquifers (< 25 m is the most effective way to alleviate salinity effects near the surface.

Tile drainage is a very effective way of controlling water tables and reducing waterlogging in areas with no pumpable aquifers. It involves the installation of slotted PVC pipe (or other material) at about 1 m below the surface. The soil water enters the pipe through the slots and is carried to a central well pit where it can then be pumped away via either irrigation or drainage systems. Tile drainage is very expensive. If the above targets are met, the whole community will feel the benefits.

7.3.3 Disposal of drainage water

Besides the benefits to the land being drained, one needs to consider the disposal of the drainage water. Several methods have been tried in some countries with complete or partial success.

The disposal of rivers is one option widely used in practice. The disposal can be controlled or uncontrolled. In the case of controlled disposal, the volume, salinity and time of disposal are controlled. Effluents from the drainage facilities are collected in holding basins or evaporation basins and released into the river during high flow periods. The common practice of uncontrolled disposal of the drainage effluent merely serves to salinize the vital fresh water supply for the less fortunate users who happen to be located downstream from the point of discharge. If they also drain their fields in a similar fashion, the river will undergo progressive salinization and its lower reaches may become unfit as a water source for either human use or irrigation. The river then turns into a saline stream, with consequent effect upon its associated aquifer and estuary, or upon the lake or bay into which the river flows. If, in addition to agricultural drainage, domestic, municipal and industrial effluents are also discharged into the river, it can become in effect an open sewer, which can endanger the entire population of the region (Hillel, 1990).

The disposal to open seas, inland seas and lakes is an attractive option, but mainly for economic reasons cannot be considered as a feasible option for areas located far from the sea. With this option, saline effluents are disposed of after being concentrated in evaporation basins to reduce their volume or without concentration. The effects of the nutrients, pesticides and chemical components of the disposed waters on the aquatic life and algae growth should be carefully investigated during the period of feasibility study and environmental impact assessment. Several countries are known to dispose saline effluents by this method. For example, Egypt and Pakistan practice the disposal of their agricultural saline effluents to the Mediterranean and the Arabian Sea, respectively.

Drainage water can be disposed to evaporation basins (ponds). These can be formed from natural/constructed depression or saline lakes (Evans, 1989). They must be designed and operated with care since drainage water with high levels of trace elements can be considered hazardous or attain toxic levels in the evaporation basins. Drainage water may be treated to remove toxic constituents before ponding, but this is costly.

Evaporation basins vary in size from a few hectares serving individual farms to very large basins (thousands of hectares) serving large areas (regional/sub-regional). Their advantages include effectiveness if they are properly designed, operated and maintained; salt harvesting potential; and the possibility of providing a habitat for a diversity of bird life. Their disadvantages consist of adverse effects due to lateral seepage causing salinity and waterlogging in adjoining areas, their finite life, their land requirements, and possible adverse environmental impacts. One of the major problems associated with the operation of evaporation basins is leakage. There are two broad types: lateral and vertical leakage.

The use of on-farm evaporation basins to dispose of saline water from tile drainage or groundwater pumping on individual farms also has several problems such as the effectiveness of basins located on the more permeable soils will be limited unless they are lined, and the potential to adversely affect neighboring properties would be higher than that of public, large-scale basins, which are usually sited on lower-permeability soils. However, the on-farm basins may be well used as short-term holding basins with release permitted during high flow periods.

Deep-well disposal by injecting brine into isolated deep aquifers can be a convenient and economically feasible disposal method in particular areas with adequate geologic and hydro-geologic characteristics. Problems, failures and environmental consequences of poor deep-well injection are mostly related to errors in the good construction, undetected pathways for fluid migration, operational errors and natural events such as earthquakes and other specific problems that may arise in the injection of brine requiring compromise (Gutteridge et al., 1983). Deep-well injection schemes should have adequate monitoring facilities to provide evidence that the injection well is operating correctly and that the injected fluids are being contained to avoid environmental problems.

7.4 Other water-related issues

7.4.1 Desalination

Desalination is the process of separating the dissolved salts content of saline or brackish water, to render it suitable for domestic, industrial or other purposes. Various desalination processes are: reverse osmosis (RO), electro-dialysis (ED), ion exchange (IE), multi-stage flash distillation (MSF), vapor compression (VC) and solar distillation. In the RO process, ions are transported through a membrane from one solution to another under the influence of a direct current electrical potential. IE involves the passage of the saline water through two beds of resins, one of which removes the anions and the other the cations. Pure water emerges from the second bed. After a time the beds become exhausted and are regenerated. In MSF, the heated saline water is passed through a series of chambers under a gradually increasing vacuum. The water thereby "flashes" into a steam-water mixture and the steam is separated and condensed to pure water. In VC, steam produced from boiling water is compressed mechanically to produce freshwater. Solar distillation in its simplest form consists of a basin with a heat-absorbing surface (usually colored black) containing the saline water. The basin is covered by a sloping transparent glass or clear plastic, which transmits sunlight and condenses vapor evolved from the saline water. The condensate trickles down the slope of the cover to the edge and is collected in a trough. All desalination processes produce, as well as freshwater, a by-product with high salt content which needs to be disposed of. Desalinization plants located on the seaboard can conveniently discharge their by-product to the sea, but for inland plants disposal is a problem. A possible solution could be disposal in evaporation ponds or salt lakes.

7.4.2 Conjunctive/cyclic use of surface and groundwater

Irrigation wells are installed primarily to provide additional water for irrigation, either as insurance against drought or as a long-term supplementary source of water to the channel-supplied surface water. There is also a growing awareness of the salinity benefits of groundwater pumping and the need for a fully integrated groundwater and surface water allocation to deal with salinity issues (Evans and Nolan, 1989).

However, groundwater extraction wells can be installed for conjunctive use of surface and groundwater to better control salinity and waterlogging. If the groundwater quality is not suitable, it can be mixed with good quality water to make it suitable for irrigation or cyclic use of good and bad quality water can be practiced. Currently the conjunctive use of surface and groundwater is practiced widely in Pakistan.

7.4.3 Reuse of drainage water

Depending on its quality, drainage water may be a total or partial substitute source for irrigation water, brackish water can sometimes be used to irrigate salt-tolerant crops. According to Rhoades (1984), drainage water can be used again for irrigation to the extent that it still has value for use by a crop of higher salt tolerance. This could be achieved by a "cyclic" drainage water reuse strategy (successive irrigations of a sequence of crops of increasing salt tolerance). The strategy is to irrigate the moderately sensitive crops (lettuce, alfalfa, etc.) in rotation with river water and the salt-tolerant crops (cotton, sugar beet, etc.) with drainage water. When the quality of drainage water deteriorates to the extent that its potential for reuse is exhausted, the drainage water should be disposed to evaporation ponds or by any other convenient means. For the salt-tolerant crops, the switch to drainage water would usually occur after seedling establishment (pre-plant irrigations and initial irrigations being made with river water).

The feasibility of this strategy is supported by the following observations: (i) maximum soil salinity in the root-zone resulting from continuous use of drainage water does not occur when such water is only used for short time; (ii) substantial alleviation of salt build-up resulting from irrigation of salt-tolerant crops with drainage water occurs during the time the salt-sensitive crops are irrigated with river water; and (iii) proper pre-plant irrigation and careful irrigation management during germination and seedling establishment leaches salts out of the seed area and from shallow soil depths. The qualities of all these crops were never inferior and were often superior when they were grown using the drainage water for irrigation (Rhoades, 1990a). There are also uncertainties associated with the long-term effects of the reuse strategy, which include the reduction of the soil infiltration capacity, the potential soil salinization, and the accumulation of certain elements (selenium, molybdenum, heavy metals, etc.) in soils and plants, which are toxic to the consumers of the crops (humans and animals). Rhoades (1990b) described the management considerations of the reuse strategy, including the criteria for crop selection. Drainage water can also be used for tree plantation and agro-forestry (Schofield, 1990). Woodlots can reduce the volume of drainage water that needs to be evaporated or ultimately disposed of by 75%. The same uncertainties, as explained above, with this agroforestry-evaporation pond approach have also been identified.

The serial biological concentration of salts (SBCS) concept proposed by Heuperman (1995) is almost similar and is drummed up in Australia that it may provide an alternative salt management option in badly salt-affected areas, especially where groundwater pumping or reuse of the groundwater in the farm irrigation system is not possible. The SBCS system involves the reuse of drainage water on progressively increasing salt-tolerant crops. Each crop is underlain by tile drain for the collection of water to be used to irrigate the next stage. Along the crop sequence, the volume of drainage water collected is reduced due to plant water use and the salinity of the drainage water increases since there is little or no salt uptaken by plants. The final effluent is contained in relatively small evaporation ponds making it feasible to consider the use of floor lining to eliminate leakage. These ponds could also be used for fish farming. The highly saline water may be collected in a series of ponds where through evaporation the salts may be harvested, and may have commercial values.

8. SALINE AGRICULTURE TECHNOLOGY

Efforts have been made towards the economic utilization of salt-affected soils and brackish groundwater with the development of the saline agriculture technology. Saline agriculture is the profitable and integrated use of genetic resources (plants, animals, fish, insects and microorganisms) and improved agricultural practices to obtain better use from saline land and saline irrigation water on a sustainable basis. It is not one simple system as one might consider. In fact it is a rich collection of possible systems for the use of saline land, involving combinations of salt-tolerant trees, shrubs, crops and livestock. The main component of this approach is the identification of salt-tolerant plants, which can successfully be grown under existing saline conditions and are capable of showing minimum depression in yield.

Saline Agriculture encompass many issues/topics such as (i) selection/breeding of productive and salt-tolerant crop species/genotypes or halophytes of agricultural significance; (ii) domestication of salt-tolerant wild plant species for economic exploitation of salt-affected lands; (iii) introduction into the cultivated crops/other plant species, of salt tolerant genes from their wild relatives through conventional breeding or genetic engineering; (iv) special agronomic practices including methods of land preparation to improve their productivity, planting, irrigation and drainage management strategies and fertilizers application; and (v) physiological studies with the following objectives: firstly, determine critical plant factors controlling yield under saline conditions; secondly, identify physiological differences between salt-tolerant and salt-sensitive genotypes with a view to develop selection criteria for salt tolerance and assist breeding programs; finally, improve yield through special treatments at critical stages during the ontogeny of plants.

8.1 Diversity in plant salt tolerance

Plant salt tolerance can be defined as the plant's capacity to endure the effects of excess salt in the medium of root growth (Maas, 1990). Many plants have developed special salt tolerant mechanisms such as salt exclusion or avoidance, immobilization within the plant, excretion and adjustment in the osmotic pressure and thus, grow in the hostile saline environment. There is, therefore, an approximate 10-fold range in the salt tolerance of agricultural crops. For example, yield production of wheat may decrease up to 25% with a salinity level of 8 dS m^{-1}, while production of barley is not affected at this salinity level (Ayers and Westcot, 1985).

The salt tolerance of a plant depends on many factors and interactions including environmental factors (soil fertility, physical soil condition, salt distribution in the profile, irrigation method, water quality and climatic conditions) and biological factors (stage of growth, varieties and rootstocks) and the actual plant response to salinity varies also with the agronomic management and specific growing conditions (Balba, 1995). Thus, the salt tolerance data cannot provide an accurate quantitative assessment of the crop yield losses from salinity and predict actual crop yield under saline conditions. Rhoades (1982) concluded that the salt tolerance data should be taken as a relative rating salinity tolerance of crops. However, data indicate that some plants are more tolerant to high salinity conditions than others. Generally most fruit trees are more salt-sensitive than vegetables, while forages are the most tolerant.

The choice of suitable crops is essential to have plant species/varieties that can produce satisfactorily under saline conditions. In addition the cropping pattern, seed placement, method of crop establishment, irrigation method and frequency, proper use of mulches and plant nutrients should also be taken into consideration while managing the saline soils.

8.2 Screening techniques for salt tolerance

The selection of appropriate screening techniques for salt tolerance is one of the most crucial steps in the selection and/or development of salt-tolerant varieties. Techniques, which can be used in the field or greenhouse, have recently been described by Aslam (2008).

The salt tolerance of a plant species/variety can be studied at germination and at later growth stages at a range of salt concentrations. At germination, the salt tolerance of a plant is studied by placing seeds in a Petri dish lined with a filter paper moistened with desirable salt concentration. The salt tolerance in plants at vegetative and maturity growth stages can be studied in soil, sand, gravel or aerated solutions, employing pots or small plots.

A large number of plant species (grasses, shrubs, trees and crops) have been studied so far. The overall picture that emerged from plant salt tolerance studies is that the maximum amount of salts that can be tolerated by salt-tolerant plants varies among species and even varieties of the same species. The salt tolerance of some of these plants may enable them to produce under saline conditions yields that are comparable to those obtained from salt-sensitive crops grown under non-saline conditions.

For studying the salt tolerance in plants, a gravel culture method is routinely used at Nuclear Institute for Agriculture and Biology (NIAB). The seeds are first grown in river sand for their germination and then seedlings of various genotypes at 2–4 leaf stage are transferred to plastic buckets or cemented tanks having Hoagland solution. Using a mixture of salts, the salinity is then increased stepwise and maintained at different EC_e levels: 0 dS m^{-1}, 5 dS m^{-1}, 10 dS m^{-1}, 15 dS m^{-1}, 20 dS m^{-1}, 25 dS m^{-1} , etc. in respective buckets/tanks. Seedlings are then allowed to grow for a few weeks and then dry shoot weights are taken. The relative reduction in dry weight of various cultivars compared with their respective controls is taken as the criterion for the salt tolerance. The data so obtained should be correlated with the salt tolerance data obtained in the soil culture and in the field.

In the soil culture technique the varieties/cultivars are grown in artificially salinized soil (pots or small plots) or natural saline fields. The plants are then allowed to grow for a due course of time. The dry matter and/or grain yields are taken as a criterion of salt tolerance compared to a control.

8.3 Crop growth/production on slightly and moderately salt-affected soils

The tolerance of a number of crops to varying salinity levels has been under investigation in numerous research centres. The pioneering work was done in the USA (Maas and Hoffman, 1977). Crop tolerance tables for different fields, forage, vegetables and tree crops were assembled and lists of crops and the salinity levels that might cause a 50% yield reduction were published. Later, Maas and Hoffman (1985) and Maas (1990) summarized crop salt tolerance data collected from various sources. All plants can tolerate salinity up to a certain level without a measurable loss in yield (threshold salinity). As a general rule, the more salt-tolerant is the crop, the higher is the threshold salinity level. It must be emphasized that the salt tolerance in cultivated crops is only appropriate for slightly or moderately salt-affected soils. Research work in breeding and genetic engineering advances hopes to develop cultivars of economical crops that can tolerate high salinity levels.

Crop yields decrease linearly as salinity increases above the threshold value as shown in the equation 1.1:

$$Y_r = 100 - s\ (EC_e - t) \qquad (1.1)$$

where Y_r is crop yield relative to the same conditions without salinity, t is the threshold salinity, s is the % linear rate of yield loss with unit EC_e (dS m^{-1}) increase above the threshold value. EC_e is the electrical conductivity of the soil saturation extract, which represents the average root-zone salinity. The expected yield of a crop grown at a specific level of salinity (EC_e) can be calculated. For example, the EC_e of a farmer field is 7 dS m^{-1} and the farmer is willing to use this piece of land for growing a potato crop, so the expected potato yield relative to non-saline conditions would be $Y_r = 100 - 12\ (7 - 1.7) = 36.4$.

The farmer should think very carefully in selecting the crops for the various levels of soil salinity in his farm. Salinity mapping at the farm level and an appropriate use of the crop salt tolerance data is the correct solution/guide for farmers to predict yield reduction/losses. Please see Maas (1990) for the salt tolerance data on a range of crops.

It should be noted that published data on crops salt tolerance do not represent the salt tolerance at germination and early seedling growth (Rhoades, 1982). Crop species, which are salt-tolerant during the late growth stages, may be quite sensitive to salinity during germination (US Salinity Laboratory Staff, 1954). Particular attention should, therefore, be given to crop salt tolerance during seed germination and seedling development because low yields are often related to failure to obtain a good plant stand. Hamdy (1990) suggested increasing plant density to overcome the adverse effect of salt on seed germination. Bernstein et al. (1974) recommended that seeds should not be placed in the upper center of the furrow where salts are concentrated. They recommended placing the seeds in double row beds; each row is located at the furrow shoulder away from the area of greatest salt accumulation. Ayers and Westcot (1985) recommended a heavy pre-planting irrigation. The establishment of crops can also be facilitated by germinating them in non-saline soil in small bags, hardening them at the seedling stage and then placing them in the saline environments. However, this practice is only possible with some row crops like cotton. The development of techniques to facilitate seed germination and early plant establishment under saline conditions would be of great benefit to saline agriculture technology.

8.4 Revegetation of highly saline soils

8.4.1 Use of kallar grass

Saline soils, which cannot be reclaimed due to variable circumstances, may be left for grazing pastures. Halophytes that can grow under very saline conditions are the best choice for such conditions. For example, kallar grass (*Leptochloa fusca*), a halophyte native to salt marshes in Pakistan, has also been found a suitable plant for forage, green manure/compost, and pulp for paper production. According to some reports, its distribution area includes Egypt, India, Sri Lanka and Australia. Kallar grass is a deep-rooted, perennial tall grass, which grows to a height of 1.5 m. It is a highly salt-tolerant plant, which has an excellent salt excretory mechanism through its leaves. As it does not retain most of the salts taken up, it remains reasonably palatable for farm animals. Kallar grass can survive at a very high salinity (EC_e) of 40 dS m^{-1}, but remains economical up to a salinity of 20 dS m^{-1}. It also tolerates sodic soils and waterlogged conditions.

Kallar grass, harbours nitrogen-fixing bacteria and obtains 60%–80% of its nitrogen requirements from the air. Therefore, it grows very well without fertilizer nitrogen application. It produces about 50 t of biomass per hectare, even when irrigated with brackish water. Although kallar grass can be propagated through seeds, the best field establishment is obtained by planting stem cuttings or root stubbles. Kallar grass can be sown at any time of the year in the plains of Pakistan but the best sowing time is March. For further information concerning kallar grass and its plantation for amelioration of salt-affected soils refer to Malik et al. (1986) and Qureshi et al. (1982).

NIAB has successfully demonstrated the potential of agricultural production systems using salt-tolerant plants, in particular kallar grass (*Leptochloa fusca*) for the saline, sodic and waterlogged soils of Pakistan (Malik et al., 1986). Kallar grass is most commonly used as a forage plant and it is now an established and profitable technology that is readily incorporated into the local farming systems. The economic returns to private farmers raising buffaloes with this grass are particularly encouraging. In some areas of Pakistan kallar grass has been grown on a large scale as a primary colonizer of highly salt-affected soils by the Saline Agriculture Development Project and elsewhere by progressive farmers.

8.4.2 Use of other highly salt-tolerant plant species

Kallar grass, however, is a summer plant that grows well under wet conditions. Therefore, with its sole cultivation, the supply of forage during winter and utilization of dryland salt-affected areas remains unattended. Such issues have also been tackled by the selection of plant species which remain productive during winter and/or capable of growth under water deficit conditions. Work is in progress on salt-tolerant winter forage species like *Agropyron, Salicornia* and *Atriplex* as possible sources of fodder from dryland salt-affected areas. Research indicates successful results growing salt-tolerant plants like *Acacia ampliceps, Acacia nilotica, Prosopis juliflora, Eucalyptus camaldulensis* and *Tamarix aphylla.*

There are many other halophytes, which can be grown economically utilizing salt-affected soils. For example, *Juncus rigidus* and *Juncus acutus* can grow in saline marshes or under irrigation with brackish water or even sea water. The culms provide fibers for high-quality paper production and the seeds have medicinal value. Mangroves provide fuel and fodder in coastal areas in many countries. *Atriplex* species are well-suited to forage production on saline soils, and have been planted on a large scale in Australia for grazing (Malcom et al., 1980). In fact, there is increasing interest in several countries for cultivating, breeding and managing selected *Atriplex* species for intensive forage production on highly salt-affected soils, with highly saline water. Malcolm (1992) gives criteria, field test methods and guidelines for the selection of shrub species suitable for forage production on saline land.

A publication on saline agriculture reports on the salt-tolerant plants suitable for economic production in developing countries. This publication contains sections describing salt-tolerant plant species that may serve as food, fuel, fodder, fibre and other products such as essential oils and pharmaceuticals. Each section also contains an extensive list of recent papers and additional information on these plants (National Research Council, 1990).

8.4.3 Use of tree species

Several tree species also have great potential for growing under salt-affected and waterlogged soil conditions and are capable of producing very attractive economic returns. In Western Australia more than 100 species of Australian woody plants have been screened for salt and waterlogging tolerance under glasshouse conditions (van der Moezel and Bell, 1990). Highly salt-tolerant tree species (e.g. *Eucalyptus occidentalis, Melaleuca halmaturorum* and *Casuarina glauca)* have great potential in saline drainage water reuse schemes. With salt and/or waterlogging tolerant tree-based land management strategies, productive use of salt-affected land and control of salinity may be reasonably achievable (Marcar, 1990; 1992). Certain native tree species, such as *Eucalyptus camaldulensis* have also been extensively examined, and salt-tolerant clones developed. Significant scope exists for intra-specific selection within species of proven commercial value, for example, *Eucalyptus globulus*.

Substantial progress has been made toward answering such basic questions as which species to plant, how to plant, where to plant and what density to plant. In deciding which tree species to establish, the following three important criteria should be addressed (Schofield, 1992): (i) adaptation is the primary criterion to be satisfied and includes adaptation to such factors as climate, soil, pests, waterlogging and salinity; (ii) knowledge of the potential water use of the species could allow trees to be selected to minimize the area required for tree planting and this factor is particularly important where there are strong reasons for maintaining land under agriculture; and (iii) possible multiple uses and economic benefits.

Successful establishment of suitable salt-tolerant trees and shrubs on saline sites requires the use of appropriate pre- and post-planting strategies for reducing site environmental stress (Marcar, 1992). Any strategy should aim to minimize the shock to transplanted seedlings or direct seeded plants by ensuring as low soil salinity and waterlogging regime as possible. This means that the time of planting needs to coincide with the period of maximum leaching. Several strategies have been shown to be effective. These include: construction of mounds, particularly double-ridge mounds; application of mulch, particularly straw, newspaper or plastics; and pre-conditioning of seedlings to salt and waterlogging.

Agroforestry systems by planting trees in combination with other cultivated vegetation are also regarded as a leading solution to dryland salinity. They have a potential role in controlling irrigation salinity (Schofield, 1992). Planting trees can significantly lower water tables, and thereby reverse the causal process of salinization and allow under storey plants to grow successfully. The multiple benefits deriving from innovative configurations of trees combined with agriculture, such as wide timber belts, are now being realized. Multiple benefits also refer to the range of potential uses of the trees other than salinity control, which may influence their selection. These include commercial tree planting for timber, pulp, firewood, fodder and other products, shelter and shade, wind and water erosion control, waterlogging control and aesthetics. Although salinity control may be the prime objective, the attractiveness and feasibility of tree planting may depend strongly on some of these other uses and benefits. Such agroforestry systems can improve the productivity of crops, pasture and livestock, whilst the trees provide other products and control the salinity.

9. FARMERS' PARTICIPATORY APPROACH AND COMMUNITY-BASED NATURAL RESOURCE MANAGEMENT

9.1 Farmers' participatory approach in salinity assessment and mapping at farm level

Most plants either fail to grow in saline soils or their growth is reduced significantly; few plants grow well on saline soils. Therefore, salinization often restricts the options for cropping in a given land area. The assessment of salinity at surface and subsoil layers at the farm level would identify different salinity zones in the farms and will help to design appropriate specific management practices, in particular to select adequate crops/cultivars for the respective salinity level in the farms. It is an essential tool for growers to get more value from each piece of farmland. Salinity maps also help farmers to understand subtle differences in soil properties across their fields, allowing them to develop more precise management zones and, ultimately, achieve potentially higher crop yields. For the assessment and mapping of the salinity at the farm and landscape level, a series of destructive and non-destructive techniques are described elsewhere in this publication. These techniques can be employed in consultation and if possible, with the participation of the farming communities. Salinity mapping at the whole farm scale is the best practice for adequate crop selection. Farmers should be aware of the relevance of salinity mapping for sustainable land use and management on their farms. Advisory technical programs should be developed to help farmers to plan and use salinity control practices on their farms.

9.2 Community-based natural resource management

Controlling salinization of land and water resources cannot be achieved by individual landholders working in isolation and taking local measures. Also, the government action in the form of isolated management programs run by different agencies is not the most effective way of tackling the complex problem. Instead, governments should develop a coordinated action plan with the participation of their agencies and the local communities. The government support for community education and their participation in the development, implementation, monitoring and cost-sharing of action plans is crucial for the effectiveness and success of the rehabilitation programs of salt-affected areas.

Education of the farming community on natural resource management is vital to increasing people's awareness and improving understanding of the salinity problems so that the adopted strategies are widely supported and acted upon. The government of Pakistan has taken a leading role in helping the farming community by starting an integrated program based on the farmer's participatory approach to the management and reclamation of saline soils. It is a partnership between production and conservation issues. The program involves whole community caring for their natural resources, government agencies, farming community, schools and interested individuals. The activities include formation of working groups, participatory planning processes, definition of the role of departmental officers, and community education. Salinity exhibitions for community education are arranged in government institutes; demonstration days at the farmer's fields are also useful. Information brochures on salinity control and management at the farm level are being prepared and their distribution to the farming community is being arranged to enhance their knowledge and understanding to tackle salinity sustainably. This community-based program has resulted in a semi-large scale rehabilitation of salt-affected soils and a substantial enhancement of the income of the participating farmers.

REFERENCES

Abrol, I. P., et al. (1988). Salt-affected soils and their management. *FAO Soils Bulletin*, *39*, 131, FAO, Rome.

Adams, W. M., & Hughes, F. M. R. (1990). Irrigation development in desert environments, In Goudie, A. S. (Eds.), *Techniques for desert reclamation*, 135-160, John Wiley, New York.

Albert, W., et al. (1987). *Employment of tensiometers for irrigation control*, Transactions 13th Int. Congress Irrigation and Drainage, International Commission on Irrigation and Drainage, vol.1B, 1061-1080, Rabat, Morocco, New Delhi.

Al-Gabri, I. M. (1997). *Salt-affected soils in Oman*, Management of Salt-Affected Soils in the Arab Gulf States, Proc. Reg. Workshop, FAO, 133, United Nations Regional Office for the Near East Region, Cairo.

Al-Ghanem, G. A. (1997). *Salt-affected soils in Qatar*, Management of Salt-Affected Soils in the Arab Gulf States, Proc. Reg. Workshop, FAO, 1-5, United Nations Regional Office for the Near East Region, Cairo.

Al-Ghawas, S. (1997). *Salt-affected soils in Kuwait*, Management of Salt-Affected Soils in the Arab Gulf States, Proc. Reg. Workshop, FAO, 132, United Nations Regional Office for the Near East Region, Cairo.

Al-Mehrizi, R. (1997). *The use of salt-affected soils and saline water in agriculture in the UAE*, Management of Salt-Affected Soils in the Arab Gulf States, Proc. Reg. Workshop, FAO, 129, United Nations Regional Office for the Near East Region, Cairo.

Amer, M. H. (1990). *Design of drainage system with special reference to Egypt*, Land Drainage for Salinity Control in Arid and Semi-Arid Regions, Proc. Symp., 377, Delta Barrage, Drainage Research Institute, Cairo.

Aslam, Z. (1985). *Growth, ion relations and carbohydrate levels of Atriplex amnicola at high salinities*, PhD. Thesis, School of Agriculture, The University of Western Australia, Australia.

Aslam, Z. (2008). Salt tolerance and screening techniques for sat tolerance in plants, Sustainable crop production on salt-affected land, Proc. Int. Conf., Saline Agriculture Research Centre, UAF, Pakistan.

Ayers, R. S., & Westcot, D. W. (1985). *Water Quality for Agriculture*, FAO Irrigation and Drainage Paper, No. 29 Rev. 1, FAO, 174, Rome.

Balba, A. M. (1995). *Management of problem soils in arid ecosystems*, Lewis Publishers, New York-London, 250.

Bernstein, L., Francois, L. E., & Clark, R. A. (1974). Interactive effects of salinity and fertility on yields of grains and vegetables. *Agronomy journal*, *66*(3), 412-421.

Boyden, S. (1987). *Western Civilization in Biological Perspective: Patterns in Biohistory*, Oxford

University Press, Oxford, 370.

Boyle Engineering Corporation. (1986). *Evaluation of On-Farm Management Alternatives*, Publication prepared for the San Joaquin Valley Drainage Program, Boyle Engineering Corporation, Fresno, California, 300.

Brown, P. L., et al. (1983). *Saline Seep Diagnosis, Control and Reclamation*, Conservation Research Report No. 30, US Department of Agriculture, Agricultural Research Service, Washington D. C., 22.

Campbell, G. S., & Campbell, M. D. (1982). Irrigation scheduling using soil moisture measurements: theory and practice, In Hillel, D. (Eds.), *Advances in irrigation*, *1*, 25-42.

Casey, H. E. (1972). Salinity problems in arid lands irrigation: a literature review and selected bibliography, University of Arizona, Office of Arid Land Studies, Tucson, Arizona, 300.

Castle, D. A., McCunnall, J., & Tring, I. M. (1984). Field drainage; principles and practices, Batsford Academic and Educational, London, 250.

Driscoll, F. G. (1989). *Groundwater and Wells* (2nd edition), Johnson Filtration Systems Inc., St. Paul, Minnesota, 1089.

Euroconsult. (1989). *Agricultural Compendium for Rural Development in the Tropics and Subtropics* (3rd edition), Elsevier, Amsterdam, 740.

Evans, R. S. (1989). Saline water disposal options in the Murray Basin. *BMR J. Australian Geology & Geophysics*, *11*(2-3), 167-185.

Evans, R. S., & Nolan, J. (1990). A groundwater management strategy for salinity mitigation in the Victorian riverine plain, Australia. *Groundwater management*, *11*, 487-499.

Food and Agriculture Organization of the United Nations (FAO). (1980). *Drainage Design Factors: 28 Questions and Answers*, FAO, Rome, 52.

Francois, L. E. (1981). Alfalfa management under saline conditions with zero leaching. *Agronomy Journal*, *73*(6), 1042-1046.

George, P. R. (1985). Sub-surface drainage methods for salinity control. *Journal of the Department of Agriculture, Western Australia, Series 4*, *26*(4), 112-114.

Ghassemi, F., Jakeman, A. J., & Nix, H. A. (1995). *Salinisation of land and water resources: human causes, extent, management and case studies*. CAB international, 526.

Greenway, H., & Munns, R. (1980). Mechanisms of salt tolerance in nonhalophytes. *Annual review of plant physiology*, *31*(1), 149-190.

Gutteridge, H., et al. (1983). *The Application of Salinity Control Techniques in Victoria*, Gutteridge Haskins & Davey, Melbourne, 131.

Hamdy, A. (1990). *Crop management under saline water irrigation*, The use of saline waters for crop production, FAO-AGL/MISC/16/90, Expert consultation on Water, Soil and Crop

Management Relating to the Use of Saline Water, FAO, Rome, 108.

Hawkins, G. P. (1978). Modern trends in mechanisation of construction of irrigation and drainage projects. State-of-the-Art Irrigation Drainage and Flood Control, In Framji, K.K. (Eds.), *International Commission on Irrigation and Drainage no. 1*, New Delhi, 107-254.

Heuperman, A. F. (1995). *Salt and water dynamics beneath a tree plantation growing on a shallow water table*, Report of the Dept. of Agriculture, Energy and Minerals, Institute for Sustainable Irrigated Agriculture, Tatura Center, Victoria, Australia, 61.

Hillel, D. (1990). Ecological aspects of land drainage for salinity control in arid and semi-arid regions. In *Proceedings of Symposium on Land Drainage for Salinity Control in Arid and Semi-arid Regions, Cairo.* (No. 1, pp. 125-135). Drainage Research Institute, WRC (Ministry of Public Works and Water Resources).

Jackson, R. D. (1982). Canopy temperature and crop water stress, In Hillel, D. (Eds.), Advances in Irrigation, Academic Press, New York, *1*, 43-85.

Maas, E. V., & Hoffman, G. J. (1977). Crop salt tolerance – current assessment. *Journal of the irrigation and drainage division*, *103*(2), 115-134.

Maas, E. V., & Hoffman, G. F. (1985). Crop salt tolerance – current assessment, Water quality for agriculture, *FAO Irrig.* Drain. Paper No.29, Rev. 1, 36.

Maas, E. V. (1990). Crop salt tolerance, Agricultural Salinity Assessment and Management, In Tanji, K. K. (Eds.), *American Society of Civil Engineers*, New York, 262-304.

Malcom, C. V. (1992). *Selecting forage shrubs for productive use of saline land*, Productive Use of Saline Land, Proc. Nat. Workshop, Adelaide Department of Agriculture, Australia, 52-58.

Malcom, C. V., et al. (1980). *Niche seeding for broad scale forage shrub establishment on saline soils*, Proc. Intl. Symp. Salt-Affected Soils, Symposium papers, CSSI, Karnal, India, 539-594.

Malik, K. A., et al. (1986). Kallar Grass: A Plant for Saline Land, *Nuclear Institute for Agriculture and Biology*, *Faizalabad*, Pakistan, 93.

Marcar, N. E. (1990). *Tree option for utilisation of salt-affected land*, Revegetation of Saline Land, Proc. Workshop, In Myers, B. A., & West, D. W. (Eds.), Institute for Irrigation and Salinity Research, Tatura, Victoria, 53-59.

Marcar, N. E. (1992). *Trees for salt-affected land*, *Productive Use of Saline Land*, Proc. Nat. Workshop, Adelaide Department of Agriculture, Australia, 12-23.

Mashali, A. M. (1995). *Network on integrated soil management for sustainable use of salt-affected soils*, Salt-Affected Lagoon Ecosystems, Proc. Int. Symp. ISSALE-95, 267-283.

McWhorter, D. B. (1977). *Drain spacing based on dynamic equilibrium*, J. Irrig. Drain. Division ASCE 103 (IR2), 259-271.

National Research Council. (1990). *Saline agriculture: salt-tolerant plants for developing countries*. National Academies Press, 143.

Oldeman, L. R., Hakkeling, R. T. A., & Sombroek, W. G. (1991). *World map of the status of human-induced soil degradation: an explanatory note* (2nd rev. ed.). International Soil Reference and Information Centre, 35.

Oster, J. D. (1984). Plant water uptake or how do plants integrate? *Soil Water*, *52*, 11-13.

Postel, S. (1989). *Water for agriculture: facing the limits*, Worldwatch Paper 93, Worldwatch Institute, Washington DC, 59.

Postel, S. (1990). *Saving water for agriculture*, State of the World 1989, A Worldwatch Institute report on progress toward a sustainable society. World Watch Institute, Washington DC, 39-58.

Qureshi R. H., et al. (1982). *Diplachne fusca*: An Australian salt-tolerant grass used in Pakistani agriculture. *J. Aust. Inst. Agric. Sci*, *48*(4), 195-199.

Rhoades, J. D. (1980). Determining leaching fraction and field measurements of soil electrical conductivity. *Agricultural Water Management*, *2*, 205-215.

Rhoades, J. D. (1982). *Reclamation and management of salt-affected soils*, Proc. 1st Annual Western Provinces Conf, 123.

Rhoades, J. D. (1984). Reusing saline drainage waters for irrigation: a strategy to reduce salt loading of rivers, Salinity in Watercourses and Reservoirs, State-of-the-Art Control of Salinity, Proc. Int. Symp. In French, R. B. (Eds.), Butterworth Publishers, Boston, 455-464.

Rhoades, J. D. (1990a). *Assessing suitability of water quality for irrigation*, Water, Soil and Crop Management Relating to the Use of Saline Water, In Kandiah, A. (Eds.), Food and Agriculture Organization of the United Nations, Rome, 52-70.

Rhoades, J. D. (1990b). *Strategies to facilitate the use of saline water for irrigation*, Water, Soil and Crop Management Relating to the Use of Saline Water, In Kandiah, A. (Eds.), Food and Agriculture Organization of the United Nations, Rome, 125-136.

Schofield, N. J. (1990). *Salinity problems and remedies (with particular reference to agroforestry) in the San Joaquin Valley*, California, Land and Water Research News 6, Western Australian Steering Committee for Research on Land Use and Water Supply, 14-17.

Schofield, N. J. (1992). Tree planting for dryland salinity control in Australia. *Agroforestry Systems*, *20* (1/2), 1-23.

Schofield, N. J., et al. (1989). Vegetation Strategies to Reduce Stream Salinities of Water Resources Catchments in South- West Western Australia, Water Authority of Western Australia, Leederville.

Shahid, S. A., & Jenkins, D. A. (1992). Utilization of a simulation system for quick screening of soils against salinity and sodicity, Advances in Planning, Design and Management of Irrigation Systems as Related to Sustainable Landuse, In Feyen, J., et al. (Eds.), *2*, 615-626.

Szabolcs, I. (1989). Salt-Affected Soils, CRC Press, Boca Raton, Florida, 274.

US Salinity Laboratory Staff. (1954). Diagnosis and Improvement of Saline and Alkali Soils, USDA Agricultural Handbook No. 60, In Richards, L. A. (Eds.), US Government Printing Office, Washington DC, 160.

US Salinity Laboratory Staff. (1981). Minimizing salts in return flow through irrigation management, Final Report, Interagency project EPA-IAG-D6-0370, *US Salinity Laboratory Staff*, US Environmental Protection Agency, Washington DC, 160.

van der Moezel, P. G., & Bell, D. T. (1990). Saltland reclamation: Selection of superior Australian tree genotypes for discharge sites. *Proc. Ecological Society of Australia, 16*, 545-549.

Zijlstra, G., & van Someren, C. L. (1980). Development in subsurface drainage techniques, Land Reclamation and Water Management: Development, Problems and Challenges, In Wiersma-Roche, M. F. L. (Eds.), *International Institute for Land Reclamation and Improvement (ILRI)*, Wageningen, 171-180.

CHAPTER 2.

LAND AND WATER SALINITY

Philip Hollington

CAZS Natural Resources, School of Natural Sciences, Bangor University, Thoday Building, Deiniol Road, Bangor LL57 2UW, Wales, UK

1. INTRODUCTION

Soil and water salinity and sodicity are major abiotic constraints affecting agriculture in dryland and irrigated areas of the world. The main purpose of this Chapter is to describe the causes and extent of the salinity and sodicity problems in soil and water and their effects upon soils as well as to describe briefly the methods of assessment of both soil and water salinity.

2. WORLD LAND AND WATER RESOURCES

This section provides a brief assessment of the current state of the world's total land and water resources, land and water use for agriculture, the importance of irrigated agriculture on global food production and its relationship to population, especially in the developing world.

2.1 Land resources

The total and potential arable land areas as well as the currently cultivated and irrigated areas in each area are shown in Table 2.1. Clearly, the potential arable land is unevenly distributed around the world, and much of it also has marginal soils or climatic conditions. Just below 50% of the potential arable land is cultivated, thus over 1,700 Mha remain to be developed–just more than it is currently used. However, the best arable land is already in use, so any increase in food production will have to come from marginal or degraded lands.

Table 2.1 Total land area, potential arable land, cultivated and irrigated land (Mha) (derived from Ghassemi et al., 1995)

Area	Land area	Potential arable land	Cultivated land	Irrigated land
Africa	2,964	734	185	11
Asia	2,679	627	451	142
Oceania	843	153	49	2
Europe	473	174	140	17
North America	2,138	465	274	26
South America	1,753	681	142	9
Ex-Union of Soviet Socialist Republics	2,227	356	233	20
Total	13,077	3,190	1,474	227

Table 2.2 shows the changes of the land resources area by categories over the period 1980–2000. Arable land worldwide increased by only 3.7% between 1980 and 1990, and of this only 0.3% was in developed countries, with a 9% increase in developing countries (FAO, 2007a).

Table 2.2 Evolution of world land resources by categories (Thousand ha) (FAO, 2007a)

Category	Year	Area	Percentage of total (%)
Arable	1980	1,345,989	10.35
	1990	1,395,973	10.73
	2000	1,397,656	10.75
Permanent crops	1980	102,020	0.78
	1990	119,883	0.92
	2000	135,821	1.04
Pasture	1980	3,244,404	24.95
	1990	3,368,403	25.90
	2000	3,442,078	26.47
Irrigated	1980	209,657	1.61
	1990	244,196	1.88
	2000	275,090	2.12
Total		13,004,202	

Available land *per capita* decreased from 0.38 to 0.28 ha (in developing countries the decline was from 0.28 to 0.20 ha per person). At the current level of use, without new land being cultivated and with none being lost, *per capita* land availability will fall to 0.14 ha in a few years.

2.2 Water resources

Freshwater (FW) is a very scarce resource in the world. It accounts for around 2.5% of the total water supply, and of it only around 30% (or less than 1% of the total water in the world) is readily available for agriculture (Table 2.3).

Table 2.3 World water resources (Derived from Table 2 in Ghassemi et al., 1995)

Category	World reserves		
	Vol (10^3×km^3)	Percentage of total (%)	Percentage of FW (%)
Oceans	1,338,000	96.5	
Fresh groundwater	10,530	0.76	30.1
Soil water	17	0.001	0.05
Glaciers & snow	24,064	1.74	68.7
Permafrost	300	0.022	0.86
Freshwater lakes	91	0.007	0.26
Salt water lakes	85	0.006	
Marsh & rivers	14	0.001	0.04
Biological water	1	0.0001	0.003
Atmosphere	13	0.001	0.04
Total water	1,385,985	100	
Freshwater (FW)	35,029	2.53	100

Note: Estimated average annual precipitation is about 800 mm, but this rainfall is not distributed evenly and many areas receive considerably less than that amount.

2.3 Global water use

By the year 2000, more than half of the available water was utilized each year, and over the course of the 20th century, global water use increased 10-fold. In 1900, agriculture used about 90.5% of the available global water, and by 2000 this was down to 70%, mainly as a result of increasing competition from industry and domestic consumers (WRI, 2008). Figure 2.1 shows the extension of land suitable for irrigation and irrigated area in the developing world in 2001.

It may be noted that there is little potential for further irrigation in Asia and the Near East & North Africa, so future increases in food production will need to come from already-irrigated land. In Latin America and particularly Sub-Saharan Africa, there are still great potentials to expand irrigation, and it is likely that this will lead to increasing problems due to salinity in these regions.

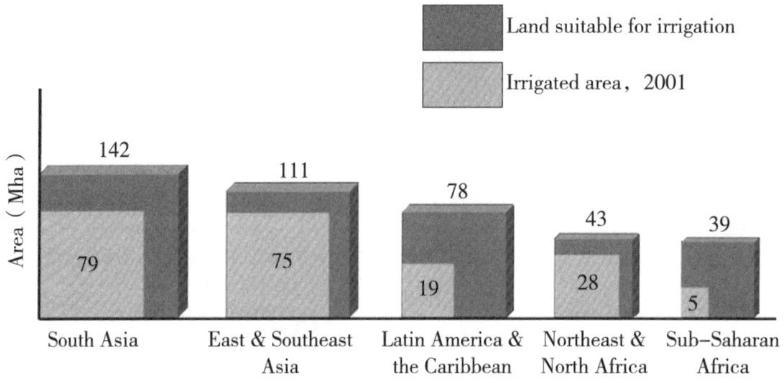

Figure 2.1 Irrigated area and land suitable for irrigation (Mha) in the developing world (FAO, 2007b).

2.4 Population growth

The world's population doubled between 1950 and 1990, and is expected to almost double again by 2030 (WRI, 2008). In 1990, 77% of people lived in the developing world (WRI, 2008), and this is expected to be over 80% by 2030. Population growth increases the demand for food, goods and services, and will result in increased competition for water. Population growth peaked at 2.0% during the period 1965–1970 and is currently at about 1.2%, although expected to fall to about 0.75% by 2030 (WRI, 2008).

A comparison of growth in irrigated land area (millions hectares, Mha) with the population (millions) in selected countries for the period 1980–2003 (Table 2.4) shows that most of the production to feed the increasing populations of the developing world has come from irrigated lands (FAO, 2007a). It is highly likely that this trend will continue in the future. This situation has two major implications: (i) with the increasing competition for water, will there be enough water available for irrigation to feed future populations? and (ii) as salinity and irrigation are inextricably linked, will irrigated lands be able to continue feeding the world in the future?

Table 2.4 Comparison of growth in population with irrigated areas in some countries (FAO, 2007a)

Country	Irrigated area (Mha)			Population (millions)		
	1980	2003	percentage of increment (%)	1980	2003	perentage of increment (%)
Bangladesh	1.5	4.7	212.5	85	147	72.6
China	45.3	54.6	20.5	1,004	1,312	30.6

(continued)

Country	Irrigated area (Mha)			Population (millions)		
	1980	2003	percentage of increment (%)	1980	2003	perentage of increment (%)
India	38.4	55.8	45.2	689	1,065	54.6
Iran	5.2	7.7	47.7	39	69	74.9
Mexico	5.0	6.3	26.9	68	103	53.1
Pakistan	14.8	18.2	23.6	81	154	90.0
Russian Federation	*	4.6	*	*	143	*
Thailand	3.0	4.9	65.8	46	63	35.6
Turkey	2.7	5.2	92.3	46	71	54.6
USA	20.6	22.4	8.8	231	294	27.0

Note: "*" indicates data are not collected.

The importance of irrigated agriculture is shown in Figure 2.2, which shows the relationship between the area irrigated and the proportion of the population undernourished in the regions of the developing world. The large-scale adoption of irrigated agriculture in South Asia has played a major role in reducing malnutrition, particularly when compared with Sub-Saharan Africa.

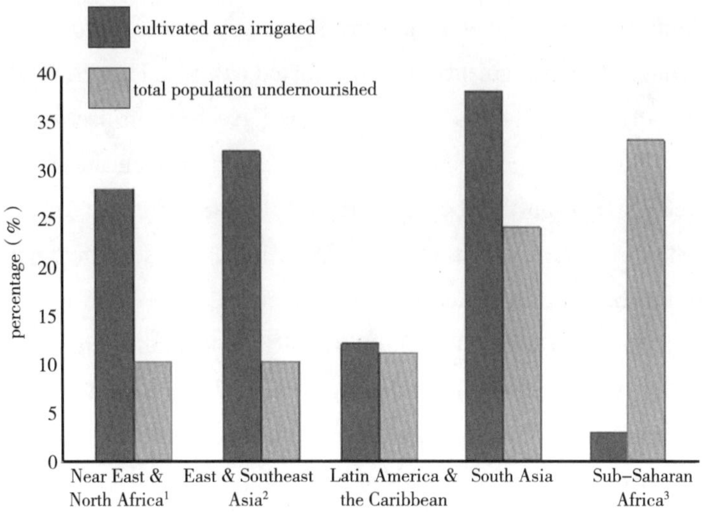

Notes: "1" indicates excluding Israel; "2" indicates excluding Japan; "3" indicates excluding South Africa.

Figure 2.2 The relationship between irrigation and malnutrition, 1998–2000 (FAO, 2007b).

Overall, the prognosis is not good. Increasing world population will require feeding from a diminishing pool of land suitable for irrigation and with increasing competition for water from domestic and industrial users. Add to this the potential effects of climate change, and it is likely that salinity, already a major constraint to production in many areas of the world, will become a more severe problem in the future.

3. HISTORICAL ASPECTS RELATED TO SALINITY

3.1 Mesopotamia

Soil salinity has been known for thousands of years. The first record of what were noticeably salinity effects was in Mesopotamia (now Iraq) 6,000 years ago (Tanji, 1990). This civilization was built upon a highly productive wheat-based irrigated agriculture in the valleys of the Tigris and Euphrates rivers, but in common with many countries since no provision was made by the engineers for any form of drainage. This led to raised water tables, and as NaCl accumulated in the soil the wheat had to be replaced by barley, a more salt-tolerant crop but less nutritious. After about 1,000 years, similar extensions of barley and wheat were grown, and by 2100 BC wheat made up less than 2% of the grain production. The result of this was that the civilization declined and eventually disappeared. Similar problems have been recorded from the early civilizations in the Americas (Tanji, 1990).

3.2 India

In India, salinity was first noted during the Vedic era around 1400 BC (Singh, 1998). At that time, lands began to be distinguished as salt-affected (*usara*). The term *kalar* also came into use about then for salt-affected land, and these terms have been in use for salt-affected lands throughout India and Pakistan since then. Salt was already a problem in Punjab and Sindh well before the large-scale irrigation that began in British India during the 19th century. However, it was only with the development of these very large irrigation schemes, without attention to drainage, that salinity become of concern to the government. In 1855 a grower near what is now the Central Soil Salinity Research Institute (CSSRI) at Karnal, India, complained about the secondary salinization following the remodelling of the Western Yamuna Canal, and the fear of the government that salinity would lead to a rejection of large-scale irrigation led them to commission a number of reports and analysises. The *Reh* (another term for salt-affected land) Committee of 1879 made a number of recommendations towards solving the problem, including mapping, and experiments to test the effects of drainage and different cultivation techniques. Work on the problem has since continued in India and, since partition in 1948, in Pakistan and later in Bangladesh.

4. CAUSES OF SOIL SALINITY

Salinity and sodicity affect a large proportion of irrigated lands throughout the world, and salt-affected soils are natural parts of arid landscapes in which evapotranspiration exceeds rainfall for a large part of the year. They are a major cause of desertification, reduced yields and increased poverty. Salinity results from the accumulation of soluble salts (excess ions) in the soil solution. The most important cations (ions with a positive charge) are Na^+, Ca^{2+}, Mg^{2+} and K^+, while major anions (ions with a negative charge) are Cl^-, SO_4^{2-}, HCO_3^-, NO_3^- and, at high pH, CO_3^{2-}. Salinity results mainly from excess Na^+ and Cl^-, and to some extent, other anions together with associated cations. Sodicity results from excess Na^+ in the soil exchange complex over Ca^{2+} and Mg^{2+}.

Salinity can be defined as either Primary salinity, which is mainly due to natural causes, or as secondary salinity, which is largely a result of poorly-managed irrigation or other human activity. Primary salinity is the result of natural processes leading to a salt accumulation over a long-term in a particular area. This could include the long-term weathering of rocks, or be the result of previous submergence in sea water. Examples are as follows.

- High saline groundwater tables, due to weathering of rocks, topography, etc.
- Saline seeps
- Pollution
- Seawater intrusion
- Blowing of salt-laden sand from coastal areas

Secondary salinity principally results from irrigation mismanagement. It is the result of the salt in the soil profile being mobilized by excess water, which either raises water tables or increases the pressure of confined aquifers, forcing saline water upwards. Once close to the soil surface, the water evaporates and leaves the salts behind. Typical causes are as follows.

- Seepage from irrigation canals
- High salinity in irrigation water
- Insufficient leaching of salts
- Types of salts in the irrigation water

In all cases, the salt exists as dissolved solids in the rainwater, within the soil profile or in the irrigation water. Rainwater can contain up to 50 mg L^{-1} of salt, and even a salt content as low as 10 mg L^{-1} can deposit 10 kg ha^{-1} of salt to soil over a year for each 100 mm of rainfall (Ghassemi et al., 1995). Over geological time, this is a huge amount. The amount of salt in the soil varies with soil type. It is generally less in sandy soils, and is less in areas of high rainfall. The salt stored in aquifers can be enormous. A 40-m thick aquifer can, for example, store up to 400 t of salt per hectare. Finally, irrigation water can also add substantial quantities of salt to the soil. Even reasonable quality water, containing 500 mg salt per litre, will add 0.5 t of salt per 1,000 m^3

of water, and irrigating with 10,000 m^3 per year will add 5 t of salt to every hectare (Ghassemi et al., 1995).

4.1 Salinity and waterlogging

Waterlogging is often but not always associated with salinity. Where it occurs it tends to be highly seasonal, particularly in areas with monsoonal climates. Waterlogging itself is detrimental to plant growth but when combined with salinity the adverse effects are greatly enhanced (Barrett-Lennard, 2003). For this reason, the drainage conditions should be recorded when sites are assessed for salinity. The combined effects of salinity and waterlogging on plant growth will be covered in detail in the next chapter.

4.2 Changes in land use

Land clearing for agriculture can alter the hydrological balance of the area, catchment or region and is a major cause of rising water tables and salinity problems. Usually there is less ET from crops and pastures than from deep-rooted, native vegetation, thus more water percolates through the soil profile and recharges the aquifers. If the aquifers are confined or semi-confined then their pressure increases, forcing water upwards towards the soil surface and leading to the salinization not just of the soil but also of the watercourses. This has been a particular problem in Australia, where large-scale clearance of native vegetation in the 19th and 20th centuries has led to severe salinity problems, including the pollution of domestic, industrial and irrigation water supplies (Ghassemi et al., 1995).

4.3 Dryland salinity

Salinity is also a major problem in many arid and semi-arid (low rainfall) areas, for example, in the southern half of Australia, and in the Great Plains in Canada and the USA. In Australia, such areas are known as seepage salinity or scalds. The latter resulting from wind or water erosion removing the topsoil and exposing either a saline or sodic subsoil. Saline seeps result from the discharge from saline groundwater, often accelerated by dryland farming, which allows water to move through saline strata below the root-zone. Saline seeps typically occur where native vegetation has been replaced by agriculture with a lower potential ET.

5. CLASSIFICATION OF SALT-AFFECTED SOILS

The basic classification system is based mainly upon pH, EC_e and ESP/SAR, as shown in Table 2.5. In addition, the physiography/topography and drainage status of the soils affect both the degree of salinization, and the ease of reclamation, while the clay mineralogy of the soil controls the effect of sodium on soil structure and permeability. The properties of the three main types of salt-affected soils are summarized in Table 2.5 below.

Table 2.5 USDA classification of soil salinity (based upon Richards, 1954)

Soil	EC_e (dS m^{-1})	ESP (%)	pH	Description
Saline	> 4	< 15	Usually < 8.5	Non-sodic soil with sufficient soluble salts to interfere with the growth of most crops
Saline-sodic	> 4	> 15	Usually < 8.5	Soils with sufficient exchangeable sodium to interfere with the growth of most plants, and also containing appreciable quantities of soluble salts
Sodic	< 4	> 15	Usually > 8.5	Soils with sufficient exchangeable sodium to interfere with the growth of most plants, but without appreciable quantities of soluble salts

In saline soils, the salt concentration is sufficient to adversely affect crop growth, i.e. there is an excess of soluble salts, usually chlorides, sulfates and bicarbonates of sodium, calcium and magnesium. In general, the EC_e of a saline soil is above 4 dS m^{-1}, with pH between 7.5 and 8.5 and the ESP is lower than 15. Very often, the soil surface has a white crust, but there may be little effect on the soil structure. The soil remains permeable, and is well drained, and adverse effects on plant growth are generally due to the osmotic effects of the excess salts.

Sodic, or black alkali soils, have high concentrations of exchangeable sodium. Typically, they have an EC_e less than 4 dS m^{-1}, pH above 8.5, and ESP above 15% (SAR above 13). The term sodic is generally preferred to alkali these days, as not all have a high pH. The high Na concentration causes soil organic matter to dissolve and this gives the soil a dark brown or black color. In these soils, the structure is deteriorated, with permeability decreased and root growth restricted. Such soils affect plant growth due to the excessive sodium on the exchange complex of the soil, which is associated with altered physical properties. The high ESP causes soil colloids to disperse, which leads to the blocking of the soil pores with consequent impeding of soil and water movement. Rainfall or irrigation water will stagnate on or near the soil surface, creating unfavorable conditions for plant growth. Such soils are very difficult to reclaim.

Saline-sodic soils have the characteristics of both saline and sodic soils, and are particularly common in Pakistan. In general, they have an EC_e above 4 dS m^{-1}, the pH is usually less than 8.5, and the ESP is usually above 15% (SAR above 13). Such soils initially have good permeability, but if attempts are made at reclamation by leaching without the use of gypsum, then the structure deteriorates rapidly and the hydraulic conductivity is reduced.

Under the early FAO classification system, salt-affected saline soils are classified as

Solonchaks, while sodic soils are *Solonetz* and other soil groups can have saline or sodic phases (Table 2.6). Heavy soils such as Vertisols and those with marked textural changes such as Planosols are more susceptible to salinization, and very often these have greater problems of reclamation and drainage. For further detailed classification of salt-affected soils using different systems, see Szabolcs (1989).

Table 2.6 FAO classification of salt-affected soils (Dudal and Purnell, 1986)

Grouping	Characterization
Solonchak	$EC_e > 15$ dS m^{-1} in the upper 1.25 m
Saline phase	EC_e 4–15 dS m^{-1} in the upper 1.0 m
Solonetz	$EC_e > 15$ dS m^{-1} in the upper 0.4 m
Sodic phase	EC_e 6–15 dS m^{-1} in the upper 1.0 m

The United States Department of Agriculture (USDA) has developed a complex soil taxonomy system which is widely used. In the soil taxonomy system, salt-affected soils are placed in great groups such as the Salorthids. Table 2.7 shows the USDA classes of saline soils according to different levels of salinity and their effects on crop growth. It is important to bear in mind that these are global classifications, and local conditions may be different in many respects. For instance, local systems for Pakistan are reported by Rafiq (1975) and Qureshi and Barrett-Lennard (1998).

Table 2.7 Soil classes according to salinity levels and effects on crop growth (Landon, 1991)

USDA soil class	Designation	EC_e (dS m^{-1})	Total salt content (%)	Crop reaction
0	Salt free	0–2	< 0.15	Salinity effects negligible, except in very sensitive plants
1	Slightly saline	4–8	0.15–0.35	Yields in many crops are restricted
2	Moderately saline	8–15	0.35–0.65	Only tolerant crops yield well
3	Strongly saline	> 15	> 0.65	Only very tolerant crops yield well

In addition to formal classification systems, systems based on indigenous knowledge and farmers' perceptions have been developed in those communities affected by the salinity problems. A typical one based upon soil color and texture is that used by the farmers in Pakistan's Punjab (Table 2.8). It also includes indicators based upon soil appearance (Table 2.9) and crop performance (Table 2.10).

Table 2.8 Farmers' classification of salt-affected soils in Pakistan's Punjab (Kielen, 1996)

Soil type	Local name
White surface, with either a good structure below the crust or hard underneath	*Chitta kalar* (white salinity)
Some patches of white crust, or the crust is very thin	*Chitta kalar*
Black appearance and with a hard upper layer	*Kala kalar* (black salinity)
Look good but have a hard layer deeper in the profile, sometimes with "stones" at about 30 cm	*Kalrathi* (hard layer) and *roor* (stones)
A lot of white salts on the surface, appear dry but are muddy underneath	Kalar shoor
Waterlogged	Sam

Table 2.9 Physical indicators of salinity and farmers' explanations for them in Pakistan's Punjab (Kielen, 1996)

Indicator	Explanation
Standing water visible 3–4 days post-irrigation	If good quality water has been used the soil has a salinity problem
Soil cracks after irrigation	If the soil has a good structure, then poor quality water has been used and the soil will eventually harden
Walking through the field sounds different before and after irrigation	Poor quality water has been used, and a flour-like layer developed on the soil surface with a hard layer below it
Walking leaves oily footprints	The soil has a salinity problem
White soil	The first sign of "white salinity" . May occur after irrigation with poor quality water, or during an extended period without irrigation
White patches	"White salinity" in some areas, either due to poor quality water or to salts coming from the soil
White soil surface	"White salinity" as above
Black soil	Severe salinity problems with major consequences for crop production
Muddy soils but with a dry-looking surface due to a white, flour-like layer	Waterlogged and very saline. Often also have "black salinity" and are very difficult for growing crops

Table 2.10 Crop indicators of salinity and farmers' explanations for them in Pakistan's Punjab (Kielen, 1996)

Indicator	Explanation
Poor germination	Salinity. Used for many different levels of salinity by all farmers, including both "white" and "black" salinity
Irregular growth	Salinity
Stunted growth	Salts deeper in the soil profile. The crop grows post-germination but the roots eventually meet the salts and in the worst cases the crop dies off
Yellow leaf burn	Too many salts in the soil burn the crop

Tables 2.9 and 2.10 also show how the farmers assess the physical effects of salinity upon soil and present several crop indicators. Farmers also have similar classifications for water quality. These indicators correlate well with existing schemes (Kielen, 1996).

6. EFFECTS OF SALINITY

Salinity has two main direct effects on cultivated plants: the decreased water availability in the soil and the toxicity effect of specific ions on particular plant processes, which leads to poor germination, stunted growth and reduced crop yields.

It is important to remark that the salt dynamics in the soil-water system are very high, thus the typical effects of salinity are very patchy, and vary over the season, particularly after irrigation and rainfall and over the years. Overall, salinization causes a series of land and water degradation issues in natural and agro-ecosystems, affecting not only agricultural production but also local industry/infrastructure and ultimately environmental sustainability and human welfare.

Table 2.11 shows the estimated annual economic damages due to secondary salinity in a number of countries, estimated in millions US dollars (M $) equivalent to 1995 currency levels. It can be inferred that, globally, the loss of income due to agricultural damages in the irrigated areas was about 11,400 million US dollars in 1995, and in the non-irrigated lands about 1,200 million. When other costs such as industry/infrastructure damages using salty water are added, the total economic damage is likely to be around 15,000 million US dollars.

Table 2.11 Estimates of economic damage due to secondary salinity (Ghassemi et al., 1995)

Country	Region	Estimated damage (M $ per year, 1995 equivalent)	Reference
Pakistan	Punjab and NWFP	300[a]	WAPDA, 1988
Australia	Murray-Darling Basin	208[a]	Murray-Darling Basin Ministerial Council, 1989
	Murray-Darling Basin	52[b]	Simmons et al., 1991
	Southwest Western Australia	50[a]	WA Legislative Assembly, 1991
	Southwest Western Australia	72[c]	WA Legislative Assembly, 1991
	Southwest Western Australia	32[d]	Water Authority of WA (Personal communication)
USA	Colorado River Basin	750[e]	Colorado River Basin Salinity Control Forum, 1993
	San Joaquin Valley, California	31[a]	El-Ashry et al., 1985
South Africa	Pretoria, Witwatersrand, Vereeniging and Sasolburg complex	29[e]	Heynike, 1981

Notes: "a" indicates Agricultural loss; "b" indicates Water supplies; "c" indicates Waterlogging; "d" indicates Stream salinity; "e" indicates Total damage.

Social impacts and associated costs are not easy to quantify, but some work has been done in India and Pakistan. In Northwest India, Joshi et al. (1996) calculated the loss of income from different crops due to salinity and waterlogging (Figure 2.3). The data showed that farm net incomes were affected by both salinity and waterlogging, and that losses due to both were particularly severe in cotton, and due to salinity in rice.

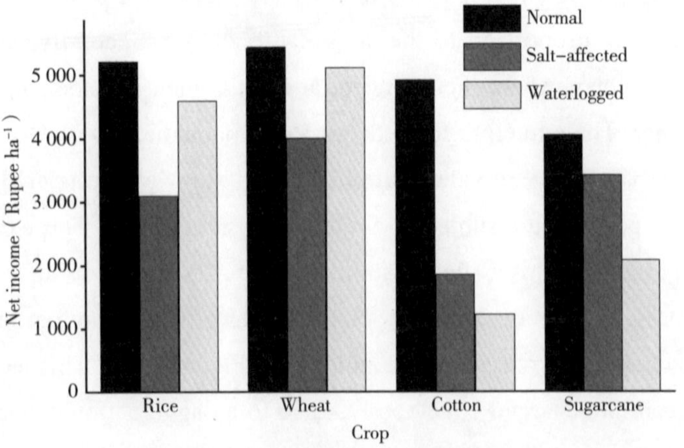

Figure 2.3 Effect of soil type on net income from various crops (Joshi et al., 1996).

These losses have effects on the entire development of the local communities, as shown in Figure 2.4 from work in Pakistan (Ijaz and Davidson, 1997). Education levels in villages located in salt-affected areas were much lower, particularly for women, and the pattern was repeated in a wide range of socio-economic indicators. Other effects may result in population movements and the abandonment of land, and may be felt on a regional level.

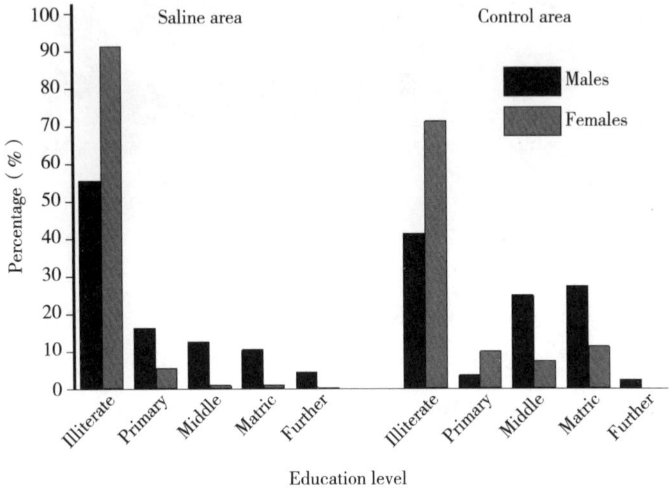

Figure 2.4 Educational achievement by gender, in saline and non-saline areas, Pakistan's Punjab (Ijaz and Davidson, 1997).

7. SOIL SODICITY

Soils become sodic when the exchange surfaces of the minute clay particles become dominated by Na^+ rather than by Ca^{2+}. The colloidal clay particles are negatively charged, and are associated with the positively-charged exchangeable cation "envelopes". Two antagonistic processes occur. Firstly, the envelope of exchangeable cations is attracted to the negatively-charged clay surface, in proportion to the size of the charge. Secondly, they tend to diffuse from the surface of the clay, where their concentration is higher, into the soil solution where the concentration is lower. The attractive force decreases exponentially with distance from the clay particle, and is twice as strong in divalent cations like Ca^{2+} or Mg^{2+}, compared to the monovalent Na^+. As a result, clay particles are subject to these electrostatic forces. These cause them to move apart (dispersion) or to come closer (flocculation) (Figure 2.5). In non-sodic (calcium-dominated) soils, the exchangeable cation envelope is compressed towards the clay particle surface because aggregative forces operate (narrow diffuse double layer, Figure 2.5B). This leads to a decrease in the repulsion between like-charged envelopes, and so to an aggregation of the clay particles. As a result, soil aggregates are stable when wet and pores in the soil are larger, and permeability and tilth improved compared to non-aggregated soils.

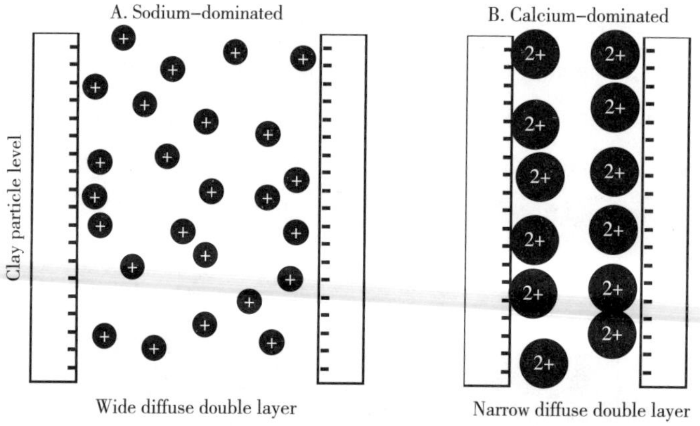

Figure 2.5 Effects of sodicity and Ca on clay particles and aggregates (Redrawn from Qureshi and Barrett-Lennard, 1998).

In sodic (sodium-dominated) soils, dispersive forces arise due to excess Na^+ (wide diffuse double layer, Figure 2.5A). The dispersive forces allow more soil solution to be imbibed between the clay particles, a phenomenon known as swelling. This reduces the size of the pores and soil permeability. This process is particularly important in soils with expandable phyllosilicate minerals such as smectites, and when the ESP > 15%. Dispersion, the release of individual clay particles from the aggregates, and slaking, the breakdown of the aggregates, can both occur below an ESP of 15% if the electrolyte concentrations are low. As a result, porosity decreases, leading to crusting, reduced permeability and a poor tilth for sowing.

If water is to enter the soil it must infiltrate through the soil surface, so the rate of entry is affected by the properties of both the water and the soil. It is, therefore, important to assess not only the salinity of the water and soil, but also the ESP and/or the equivalent SAR in the surface soil.

7.1 Use of gypsum

Gypsum is widely used as amendment to improve sodic soils. When the clay particle surfaces become dominated by Na^+, they have relatively wide diffuse double layers. However, if calcium in the form of gypsum is added, the particles become surrounded by Ca^{2+}, so the diffuse double layers narrow and the soil re-aggregates. Gypsum can either be applied to the soil or added to the irrigation water.

7.2 Effects of soil sodicity

The deterioration of the soil's physical conditions that results from the sodicity leads to a number of problems. Less air and water enter the soil, and in particular the root-zone, thus creating a poor environment for crop growth. Waterlogging develops due to poor drainage, causing seeds to rot and roots to die. Surface crusting as a result of the slaking of the aggregates leads to reduced seedling emergence, and the compaction that results from the breakdown of the soil structure creates problems for root penetration and cultivation.

8. EXTENT OF SALT-AFFECTED SOILS

Estimates of the global extent of soil salinization vary widely due to lack of consistency in the methodologies and criteria used. However, salt-affected soils are distributed all over the world, and they are a relevant problem in Asia and Australia. There are also severe salinity problems in Africa, some parts of the USA and Canada, and some European countries, for example, Spain and Hungary. Globally, over 955 Mha of area are affected by primary salinity (Szabolcs, 1989) and about 76 Mha of area by secondary salinity (Oldeman et al., 1991).

The global land area affected by salinity or sodicity and their distribution by continents is shown in Table 2.12 (Szabolcs, 1989).

Oldeman et al. (1991) assessed the the severity of land degradation processes by different causes into four categories: (i) light degradation has slightly reduced productivity and it can be manageable by local farming systems; (ii) moderate degradation has greatly reduced productivity, with major improvements needed, which generally have to be provided from outside; (iii) strong degradation implies soils that are no longer workable, and need major engineering work or international assistance for restoration; and (iv) extremely degraded soils are considered beyond reclamation. Table 2.13 presents the extent of these categories by continent as they apply to human-induced salinity while Table 2.14 shows the estimated amount of irrigated land that is affected by secondary salinization.

Table 2.12 Global distribution of salt-affected soils (Szabolcs, 1989) (Mha)

Area	Saline	Sodic	Total
North America	6.2	9.6	15.8
Central America	2.0	-	2.0
South America	69.4	59.6	129.0
Africa	53.5	27.0	80.5
South Asia	83.3	1.8	85.1
North & Central Asia	91.6	120.1	211.7
Southeast Asia	20.0	-	20.0
Europe	7.8	22.9	30.7
Australasia	7.4	340.0	347.4
Total	351.4	581.0	932.4

Table 2.13 Global extent of human-induced salinization (Oldeman et al., 1991 cited in Ghassemi et al., 1995) (Mha)

Area	Light	Moderate	Strong	Extremely	Total
Africa	4.7	7.7	2.4	-	14.8
Asia	26.8	8.5	17.0	0.4	52.7
South America	1.8	0.3	-	-	2.1
North & Central America	0.3	1.5	0.5	-	2.3
Europe	1.0	2.3	0.5	-	3.8
Australasia	-	0.5	-	0.4	0.9
Total	34.6	20.8	20.4	0.8	76.6

Table 2.14 Irrigated land in selected countries affected by secondary salinity (FAO, 2007c; Ghassemi et al., 1995)

Country	Cultivated area (Mha)	Irrigated area (Mha)	Percentage of irrigated to cropped land(%)	Salt-affected irrigated land (Mha)	Percentage of irrigated land salt-affected(%)
China	156.3	53.82	34.8	6.70	15.0
India	169.6	57.28	33.8	7.00	16.6
Kazakhstan	22.5	3.56	13.3	3.70	18.1
USA	177.2	21.40	12.0	4.16	23.0
Pakistan	22.1	17.82	80.0	4.22	26.2
Iran	18.1	6.91	38.8	1.72	30.0
Thailand	17.8	5.00	25.5	0.40	10.0
Egypt	3.5	3.42	99.9	0.88	33.0
Australia	49.7	2.54	5.1	0.16	8.7
Argentina	29.5	1.55	5.4	0.58	33.7
South Africa	15.7	1.49	9.5	0.10	8.9
Subtotal	842.8	158.70	18.8	29.62	20.0
World	1473.7	227.11	15.4	45.40	20.0

China has about 20 Mha of salt-affected soils, which are widely distributed across the territory and from them around 20% are alkaline (sodic). This extension accounts for about 21% of the arid area of China (Dregne and Chou, 1992). The country has been divided into eight salinization regions, and 27 subregions, according to the natural characteristics of salt formation, migration and accumulation in the soils, and the soil geochemistry. The eight regions are as follows.

- Coastal humid and semi-humid regions with seawater submergence
- Northeastern semi-humid and semi-arid region
- Huang-Huai-Hai Plain
- Mongolian Plateau, arid and semi-desert
- Middle and upper Yellow River, semi-arid and semi-desert
- Gansu-Xinjiang desert
- Qinghai-Xinjiang extremely arid desert
- Xizang (Tibetan) Plateau alpine frozen desert

Sodic soils are widely distributed from Inner Mongolia in the North to the Huang-Huai-Hai Plain in the South and from the East Coast to Xinjiang in the West, including the Tibetan Plateau. Saline soils are widespread in coastal areas. Sodic and saline soils often coexist. Naturally, problem areas for salinity are the gravel soils in Xinjiang and the Tibetan Plateau (Dregne et al., 1996).

China is suffering from increasing salinity problems as new irrigation schemes are established and older ones are rehabilitated. There is no whole-China data for secondary salinization, although in 1965 about 6.7 Mha were reported to be affected (Vermeer, 1977). More recently, it is reported that just under 5 Mha have been reclaimed (James, 1989). Major problem areas in irrigated lands are the Ningxia and Hetao irrigated plains along the Yellow River, and there are many areas subject to secondary salinization in the Northeast, the North China Plain, the Hexi Corridor in Gansu province, and the oases in the desert in Xinjiang.

9. MEASUREMENT AND ASSESSMENT OF SALINITY

The problem with measuring salinity in the field is that salinity changes with both space and time. A large number of edaphic (soil), climate and management factors, all of which are constantly changing and interactions between them influence the salt dynamics of a given area. The edaphic factors include landscape, topography, soil physical and chemical properties, soil parent material, and geo-hydrology of the area, especially water table depth and salinity of the perched groundwater. Management factors are cropping systems, irrigation, drainage, tillage, and agronomic practices while climatic factors include the rainfall amount and distribution, temperature, relative humidity, and wind speed and direction. As a result, a large number of samples are needed to characterize a particular field, and the measurements must be constantly updated as conditions change, or to monitor them if they are changing. Salinity assessment using traditional methods such as the laboratory analysis of soil and water samples is, therefore, time-consuming, labour-intensive and expensive. Due to the field variability, soils must be sampled when analysis allows useful predictions to be made about likely effects of salinity on plant growth. If we are conducting an experiment, for example, a screening trial of plant genotypes (G) for salt tolerance, then intensive sampling is needed over the whole season, particularly after any irrigation or rainfall, and in a large number of the experimental plots, in order to identify the environmental (E) component of the G $\times$ E interactions. However, in an agricultural field situation, then sampling mid-way through the growing season, when the soil is moist but not freshly leached, is a good compromise. In fields with high variability, samples should be taken from strongly, moderately, and slightly salt-affected areas. Depending upon the crop, sampling depth increments should be chosen as appropriate to ensure that the whole of the root-zone is sampled.

In many cases, the sampling will require considerable time and effort in the field, although very often a number of samples will be composited together for analysis. Otherwise, the chemical analysis of soil samples for salinity characterization may become very expensive.

Other methods using electromagnetic (EM) induction and advanced software for data collection are now available, including various EM techniques and time-domain reflectometry (TDR). Although these methods are far more efficient and cheaper once the initial outlay is overcome, they still demand a good calibration of the instruments with actual samples from the field. Moreover, the analysis of the physical characteristics of the soil, for example, texture, structure, mottling and gleying, root distribution, etc., is still important.

The new methods are practical, cost-effective, and faster and provide more informative outcomes: (i) they have the potential to identify the causes of salinization; (ii) they allow the salinity survey to be integrated automatically with remote sensing and GIS; and (iii) they can also identify mitigation needs.

9.1 Determination of soil salinity from electrical conductivity

Salinity refers to the accumulation of soluble salts, i.e. presence of one or more of a number of major inorganic ions (Na^+, Mg^{2+}, Ca^{2+}, K^+, SO_4^{2-}, HCO_3^-, NO_3^- and CO_3^{2-}) in the soil solution. In soils, it refers to both the soluble and the readily dissolvable salts present either in the soil or in an aqueous extract of a soil sample. Salinity can be measured either in terms of the total salt concentration or in terms of the electrical conductivity, which is generally an easier option. These two measurements are very closely related to each other, even though the electrical conductivity is also affected by the temperature and the particular ions involved.

The best measure of soil salinity is the electrical conductivity of the soil water. However, it is not a practical measure, thus the measurement is made on an aqueous extract of a soil sample, typically a saturated soil paste. Both the water saturation percentage (SP) and the water : soil ratio varies with the soil texture and they are related to the soil water content in the field. As a result of this, a standard measurement is made using the saturated soil paste extract, or saturation extract (EC_e). Conductivity can also be measured as the conductivity of the bulk soil (EC_a). Along the same lines, the salinity of the irrigation water is measured as its conductivity EC_w.

Electrical conductivity is easy to measure and provides a good index of the total concentration of ionized solutes in the soil solution. Using the SI system units, electrical conductivity is reported in siemens per metre (S m^{-1}), or for smaller amounts as decisiemens per metre (dS m^{-1}). Decisiemens per metre are equivalent to the older units of millimhos per centimetre (mmho cm^{-1}). It is important to be aware that salinity work is bedevilled by a plethora of units, some related and some not. It is recommended to stick to using the SI system units, i.e. dS m^{-1} for any field-based work, although in the laboratory and for some plant physiology, it is permissible to use units of concentration.

As conductivity is the reciprocal of the resistance, the measurement of the resistance of the soil or of the soil solution is needed. Resistance is inversely proportional to the area, and directly proportional to the length, so the resistance depends to a large extent upon the size of the sample we are measuring, and the distance between the electrodes we are using to do the measurement. The procedure to determine EC_e using the saturated paste extract is straightforward.

To prepare the saturated soil paste:

- Add distilled water slowly to 200–400 g air-dried soil, stirring all the time. The soil will reach a glistening, flowing texture that fills-in quickly after a spatula is passed through it and at this stage, it has become saturated
- Allow to stand for several hours, and overnight if possible. This allows the soil to fully imbibe water and the salts to dissolve
- Use a funnel and filter-paper, or a vacuum extractor, to obtain the extract
- Measure EC_e and temperature using standard equipment (conductimeter)
- Correct for temperature and convert to EC_e

The adjustment for temperature is made by standardizing the measurement to 25℃. This needs to be done as conductivity increases with temperature. For practical purposes, this is done by using a temperature coefficient.

$$f_t = 1 - 0.20346\ (T) + 0.03822\ (T^2) - 0.00555\ (T^3) \tag{2.1}$$

T in Equation 1.1 is calculated using the following equation :

$$T = (\text{temperature in degrees Celsius} - 25)\ /\ 10 \tag{2.2}$$

From this, the conductivity at 25℃ (EC_{25}) is calculated as:

$$EC_{25} = f_t \times EC_t \tag{2.3}$$

when EC_t is the EC_e at the measured temperature T, the EC_{25} thus calculated is the EC_e.

Conductivity can be also measured on a range of paste extracts. For instance, in a 1:5 ($EC_{1:5}$) or 1:1 soil – water ratios. Values obtained using these ratios can be converted to EC_e, but they are less reliable and are subject to error. Example formulae which can be used for rough estimates are Equation 2.4 and Equation 2.5.

$$EC_e = 2.2 \times EC_{1:1}\ (\text{a general value}) \tag{2.4}$$

$$EC_e = 6.4 \times EC_{1:5}\ (\text{Loveday, 1972}) \tag{2.5}$$

They are also less well-related to meaningful soil properties and the formulae developed tend to be location-specific. If you do want a quicker method, then the EC_e of the saturated soil paste (EC_p) as described in Rhoades et al. (1999) is the best method.

Once the EC_e of the soil has been determined, then other analyses should be carried out, in particular the concentration of individual ions. From these, such parameters as ESP and SAR can be calculated.

Some other rule-of-thumb conversions, all at 25℃, are given below.

- Total ion (cation OR anion) concentration (mmol L^{-1}) is approx. 10 $\times EC_{25}$ (dS m^{-1})
- Total dissolved solids (TDS) (mg L^{-1}) is approx. 640 $\times EC_{25}$
- Osmotic potential (MPa) is approx. 0.04 $\times EC_{25}$

9.2 Determination of field salinity from bulk soil electrical conductivity

As an alternative to the saturated soil paste extract, which is laborious and impractical on a large scale, salinity can be assessed either through direct measurements of the electrical conductivity of a saturated soil paste (EC_p), without the need to collect extracts, or the electrical conductivity of the bulk soil (EC_a). The first is a laboratory measurement or obtained using simple field equipment, while EC_a is measured either by electrodes in contact with soil or electromagnetically. From these measurements of EC_a and EC_p, then the EC_e can be obtained easily. The equations for this are given in Rhoades et al. (1999) and are not repeated here—corrections for soil physical conditions and organic matter content are needed.

The ideal situation is to know the "real time" concentrations of all the solutes in the soil water over a range of field water content. This is not yet possible, but the relatively new EM techniques enable total solutes to be easily determined and the determination of particular ions can be done in the laboratory if needed as a complement to the calibration process. The sampling strategies and sampling positions needed for these analyses can now be easily determined using computer software.

Three methods, all of which have been proved in the field, are available. These are the four-electrode probes, EM induction sensors, and TDR. The latter, however, has not yet been shown to be sufficiently accurate, simple, robust or fast enough for everyday use, and is still very expensive.

These techniques involve portable instruments that can rapidly measure bulk EC_a directly as a function of the landscape position, together with procedures and software programs to infer salinity (EC_e) directly from the EC_a prior instruments calibration. The use of appropriate software has several advantages: (i) computer-assisted mapping of the field salinity; (ii) spatial statistics analysis to infer root-zone salinity distribution; (iii) monitor salinity changes with space and time; (iv) linkage with GIS and remote sensing; and (v) the information obtained can be used to help determine the underlying causes of salinity problems, to develop solute transport models, and to predict the consequences of alternative management options.

9.2.1 Four-electrode sensors

This method is covered in detail in Rhoades et al. (1999) and, as the name would suggest, the instrument makes use of four separate electrodes inserted in the soil. The instrument also includes a current-generator and resistance meter, and a connecting wire. The depth of sampling and the volume of measurement can be altered by changing the spacing of the electrodes. Basically, when the distance between the outer pair of electrodes (the "current" electrodes) is small, then the flow of current is shallower than when they are further apart. The depth of measurement being about one-third of the distance between them (Figure 2.6). To calculate the EC_a, we need to know the spacing between the "current" electrodes and the inner or "potential" electrodes. A typical instrument is the one produced by Martek. This was specifically designed for salinity use and provides readings directly in conductivity, corrected to 25℃, rather than in resistance as some other models do.

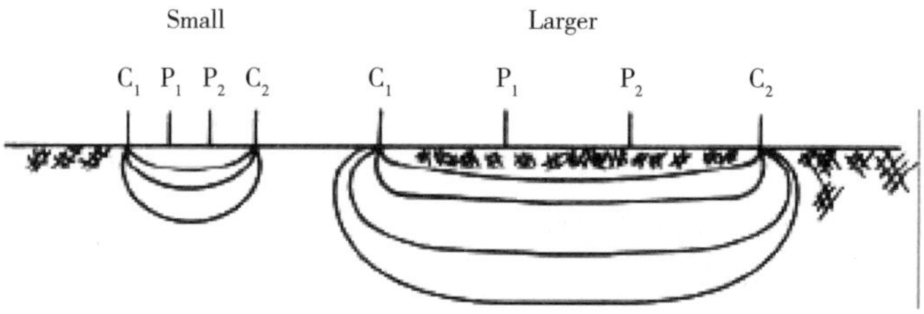

Figure 2.6 Effect of electrode spacing on depth and volume sampled (Fig. 32 from Rhoades et al., 1999).

Four-electrode probes can be mounted together on a board, so that all the electrodes can be inserted in the soil at once, to save time and make the instrument more portable in the field. Furthermore, they can be tractor-mounted and connected to data-loggers, and can also be connected to a GPS to record positions accurately.

9.2.2 Electromagnetic (EM) induction sensors

Electrical conductivity can be measured remotely by EM induction. A transmission coil at one end of the instrument induces circular, eddy-current loops in the soil, which are proportional to the conductivity. In turn, these loops generate secondary magnetic fields proportional to the loop current, which are partly intercepted by a receiver coil at the opposite end of the instrument. The value recorded by the instrument is known as the "depth-weighted" electrical conductivity or EC_a*.

The most common instrument in use is the Geonics EM38. Figure 2.7 shows a simplified diagram of the instrument where: Tx is the transmitter coil, Rx is the receiver coil, Hp is the primary magnetic field, and Hs is the secondary magnetic field. It works through a transmitter coil at the rear, which generates a primary magnetic field. This induces small eddy currents (the secondary magnetic field) in the ground. A receiving coil at the front of the instrument measures both the strong primary field and the much weaker secondary field arising from the eddy currents, which is proportional to the soil electrical conductivity.

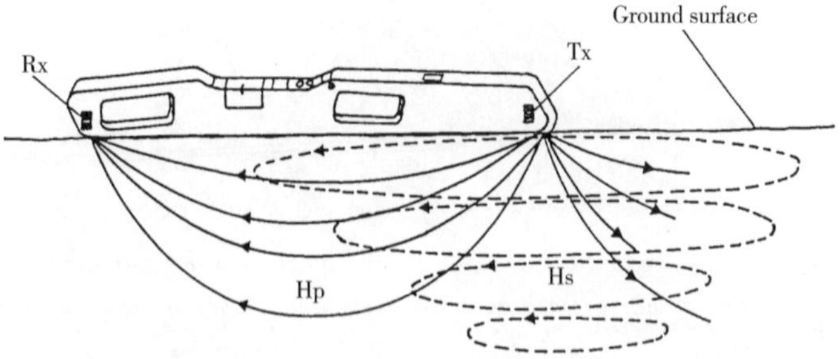

Figure 2.7 Simplified diagram of the nature of the electromagnetic fields created by the EM38 (Fig. 3 in Norman, 1990, © State of Victoria, Department of Primary Industries).

The EM38 can also be used to estimate the shape of the salinity profile within the root-zone, as it has different depth responses when placed upright on the soil surface (the vertical dipole EM_v) and when it is lying on its side (the horizontal dipole EM_h) (Figure 2.8).

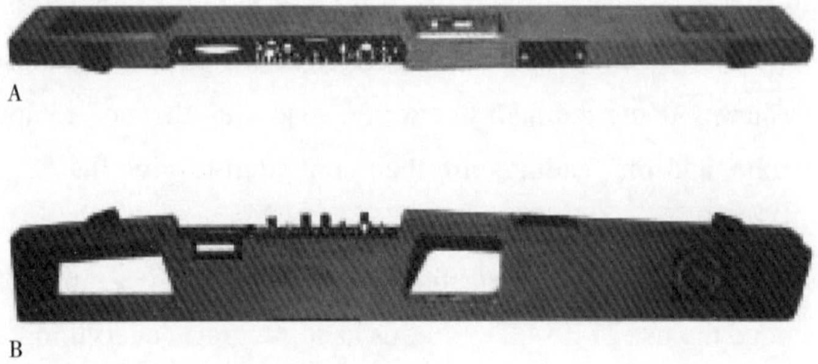

Figure 2.8 EM38 in horizontal (A, EM_h) and vertical (B, EM_v) orientation (Rhoades et al., 1999).

If the vertical reading is greater than the horizontal, then soil conductivity is greater at depth than in the shallow root-zone, and the profile is said to be leached. Conversely, if the horizontal reading is greater than the vertical, then the profile is said to be inverted. These profile descriptions may change shape over the year as a result of alternating leaching and capillarity cycles. The vertical configuration allows readings to 2 m depth, while the horizontal configuration allows readings to 1 m depth. Figure 2.9 shows the instrument in the field in Pakistan in its horizontal position.

Other EM units are available for deeper readings, and as with the 4-electrode probe, the system can be mobilized, combined with GPS or other sensors, and automated. There has also been some work to assess the possibilities of using it from an aircraft or helicopter, and this shows promise for large-scale surveys (George and Green, 2000).

Figure 2.9 An EM38 in the field in Pakistan. Note that the instrument was not actually in use at this moment, and so the demonstrator's (Dr. Ramon Aragüés) watch had not been removed!

To conduct a salinity survey using the EM38 requires calibration of the instrument to transform the depth-weighted EC_a readings to either EC_e or $EC_{1:5}$, using the vertical (EM_v) and horizontal (EM_h) readings of the EM38. It is vital to ensure that the calibrations remain accurate over time. The number of samples required for calibration depends upon the degree of variation within the field measurements, and ideally should result in a coefficient of variation for the regression of less than about 25%. The sampling pattern and number of samples required can easily be obtained using the ESAP software from the US Salinity Laboratory (2006). As noted earlier, depending upon the crop, sampling depth increments should be chosen as appropriate. When carrying out a survey using the instrument, it is also vital that soil temperature is recorded using an insertion probe, and the readings are then converted to give the EC_a at 25℃ using the relationship given above.

Once the readings have been obtained, they can be plotted by a range of methods (Figure 2.10). It is recommended the use of ESAP to do this as it integrates everything into one package.

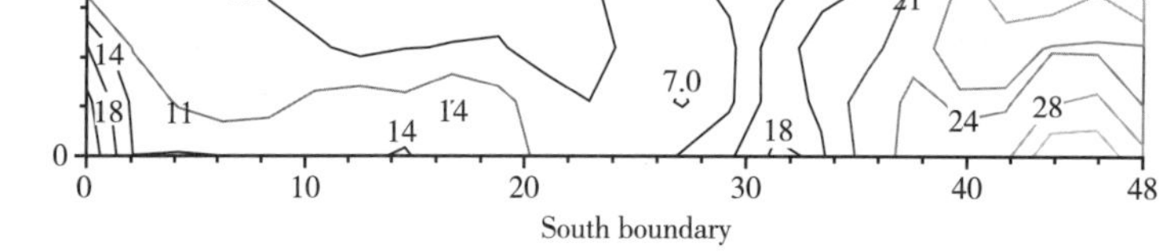

Note: ①Former name, now renamed as Khyber Pakhtunkhwa Province.

Figure 2.10 Example of the type of plot produced of salinity levels in the field [Gundheri, NWFP (Northwest Frontier Province)① *Pakistan, 13 November 1997]. The plotted data are EC_e* (ds m^{-1})*, produced from the EC_a after calibration of the EM38.*

9.2.3 Important considerations

In most field situations, the topsoil is too dry to use for sampling and assessment. Electrical currents need water and ions for transmission through the soil, and dry soil is an insulator. So no useful information about salinity or other soil properties can be inferred from EC_a measurements on the surface layers of the soil. This is a particular problem in dryland situations, but also possible on irrigated land, where very often the surface soil is highly salinized due to evaporation.

The instrument is provided with a comprehensive set of instructions (manual), and these must be read and followed carefully, in particular to zeroing the instrument and calibration. It is also important to note that, if the reading for EM_h is less than half that for EM_v, then either there is an error with the instrument, or there is some form of electrical or magnetic interference occurring.

It may be useful to use a block of wood, grooved to hold the instrument, to support the EM38 10 cm above ground level. Some researchers have found that this gives more accurate readings.

Operators should develop a routine to carry out accurate sampling:

- NEVER include the surface mulch in the soil sample for calibration
- ALWAYS scrape away dry surface soil before taking an EM38 reading
- ALWAYS record the soil temperature at regular intervals to allow the readings to be standardized
- ALWAYS check you have no metal objects on your person, any metal (watches, rings, some coins) will affect the working of the instrument and lead to incorrect readings
- ALWAYS sample away from metal poles, pylons and overhead lines, also buried electric cables and drains if possible. Water conducts electricity

9.3 Measuring soil and water sodicity

The ESP, which is the percentage of the total exchangeable charge neutralized by Na^+, is commonly used to assess the sodicity of the soil. It is calculated by dividing the soil exchangeable Na by the total soil cation exchange capacity (CEC), taking care to calculate the CEC accurately.

$$\text{ESP (\%)} = \text{exchangeable Na} / \text{CEC} \times 100 \tag{2.6}$$

It is important when assessing soil sodicity that the correct quantity of exchangeable Na is determined. To do this, the quantity of soluble Na measured from the saturation extract must be subtracted from the total extractable Na using the chemical reagent (displacement solution):

$$\text{Exchangeable Na} = \text{Total extractable Na} - \text{Soluble Na} \tag{2.7}$$

An ESP level of 15% has been proposed as the arbitrary boundary between sodic and non-sodic soils, although some workers have proposed that the level should be at an ESP of 13%, 10 or even 6. In general, however, it is safe to regard any soil with an exchangeable Na content greater than 1 mEq/100 g as potentially sodic. The effect of exchangeable Na on crops is covered in CHAPTER 4.

Another important parameter for assessing sodicity is the SAR of the soil solution or irrigation water, which reflects the balance between Na^+ and the divalent cations (Ca^{2+} + Mg^{2+}) as follows:

$$SAR = \frac{(\mathrm{Na}^+)}{\sqrt{\frac{(\mathrm{Ca}^{2+}) + (\mathrm{Mg}^{2+})}{2}}} \quad (2.8)$$

Where (Na^+), (Ca^{2+}) and (Mg^{2+}) are the ion concentrations in mEq L^{-1} of solution. The units are important because if other units are used then SAR assumes a different value.

SAR values are used to estimate water quality (see CHAPTER 2, Section 10 Irrigation Water Quality) and also to calculate the equilibrium values for ESP in irrigated soils.

For soils in equilibrium with applied irrigation water, the following equation provides a good correlation between SAR of irrigation water with measured ESP (Richards, 1954) for total cation concentrations between 39 and 110 mEq L^{-1}. This is also shown diagrammatically in Figure 2.11.

$$\mathrm{ESP} = [100 \times (0.01475\ \mathrm{SAR} - 0.0126)] / (0.01475\ \mathrm{SAR} + 0.9874) \quad (2.9)$$

Although this equation is still acceptable for most of the irrigation water used, it is now not recommended for all situations (Ayers and Westcot, 1985). When the waters contain appreciable contents of calcium and bicarbonates other equations should be used. An adjusted SAR (adjSAR) was proposed by Ayers and Westcot (1976) to take account of high carbonate waters, but this is no longer recommended as it is thought to overstate the sodium hazard.

The recommended method to estimate the sodium hazard index is now the adjusted RNa (adjRNa) procedure of Suarez (1981). This takes account of the effects of CO_2, HCO_3^- and calcium in the irrigation water, and assumes a soil source of calcium in addition. The calculated adjRNa can be substituted for SAR in any equations where that term appears. The equation to calculate adjRNa is given by

$$\mathrm{adjRNa} = \frac{(\mathrm{Na}^+)}{\sqrt{\frac{(\mathrm{Ca}_x^{2+}) + (\mathrm{Mg}^{2+})}{2}}} \quad (2.10)$$

where Na is the sodium in the irrigation water (mEq L^{-1}), Ca_x is a modified calcium value obtained from Table 2.11 in Ayers and Westcot (1985), and Mg is the magnesium in the irrigation water (mEq L^{-1}).

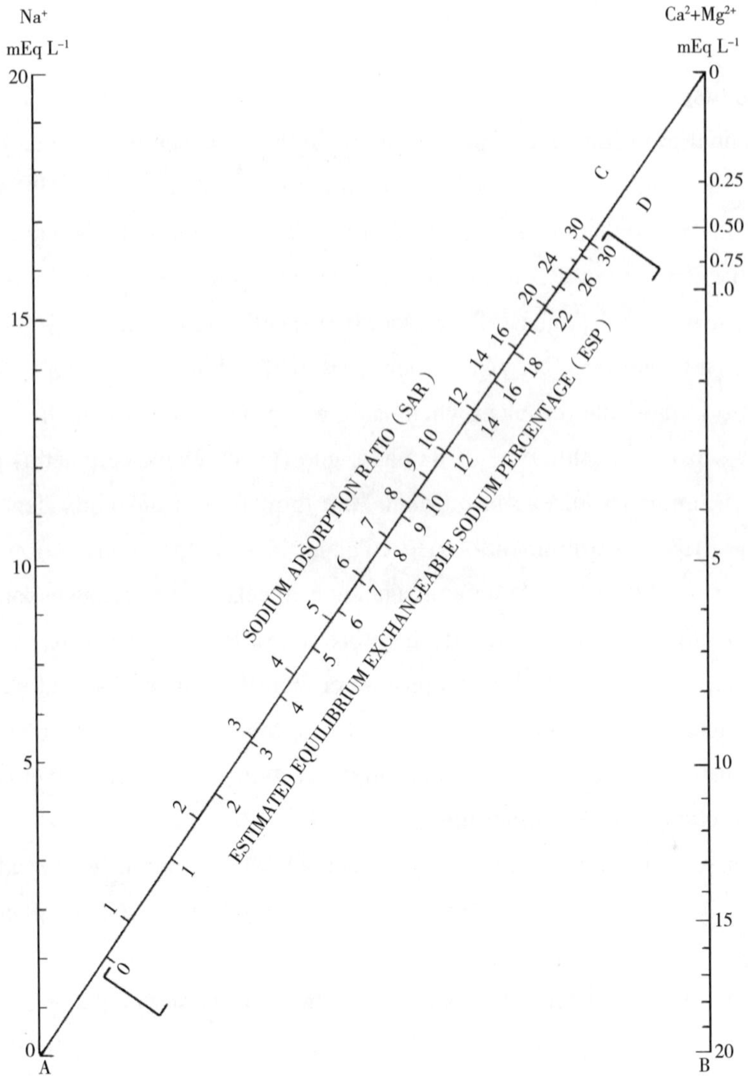

Figure 2.11 Nomogram to determine SAR of irrigation water and to estimate corresponding soil ESP (Richards, 1954).

10. IRRIGATION WATER QUALITY

Water quality for crop production is an extensive but important topic and will only be outlined here. For more detailed information, refer to Ayers and Westcot (1985) and Rhoades et al. (1992).

In general, salts are present in irrigation water only in small amounts, but they can have major effects on the soils and crops they are applied to. These effects depend not only on the concentration of the salts but also on what the salts are and their relative proportions. On the whole, a long-term view should be taken of the potential problems when determining water quality for irrigation.

Main potential problems related to restrictions in their use can be grouped into four areas: (i) irrigation water salinity, where the salts in the water or deposited in the soil reduce the water availability to the crop so that yield is affected; (ii) water infiltration rate is reduced by high Na^+ or low Ca^{2+} in the soil or the water to the extent that the amount of water infiltrating into the soil is not sufficient to supply the needs of the crop; (iii) specific ion toxicity may occur due to the accumulation of particular ions [Na^+, Cl^- or borate (BO_3^{3-})] and their effects on yields of sensitive crops. This can be a particular problem when using overhead sprinkler irrigation, and tree crops are particularly sensitive to this type of damage; and (iv) other miscellaneous problems may arise from the accumulation of excess nutrients, the formation of unsightly deposits, and from corrosion of the irrigation equipment. A particular problem is the deterioration of pumps and wells, but in some cases the concrete lining irrigation channels can be seriously damaged as well.

The main factors influencing quality here are the salinity of the water, and the ratio of sodium to the sum of Ca^{2+} + Mg^2. High salinity water increases infiltration, but water with either low salinity, or with a high ratio of sodium to Ca^{2+} + Mg^{2+}, will decrease it. If this problem persists, then crusting of seedbeds may occur together with waterlogging, increase in disease-carrying insect vectors, and weed infestations.

Salinity classes for irrigation water, equivalent salt concentration (electrical conductivity, total dissolved solids or chloride content) and their potential effects on soil and plant are shown in Table 2.15.

A sodium hazard in the irrigation water (see sodium classes in Table 2.16) can have a negative effect on crop production not by impairing the uptake of water by the crop but by reducing the infiltration rate of water into the soil and so making it unavailable to the crop. The sodicity hazard potential of water is usually assessed in terms of its SAR and its salinity levels (see CHAPTER2, section 9.3 Measuring soil and water sodicity). At the same SAR, the potential of the water to damage soil structure through dispersing the soil aggregates is greater at low salinity than at higher salinity. However, there are many interactions with other soil properties, and it is always best to directly test the effect of your water on the soil of interest.

Table 2.15 USDA classification of irrigation water salinity (Richards, 1954)

Salinity class	Description	EC_e range ($dS\ m^{-1}$)	Equivalent salt concentration (approx.)		
			TDS ($g\ L^{-1}$)	TDS ($mg\ kg^{-1}$)	Cl ($mg\ kg^{-1}$)
C1	Low salinity water. Can be used for irrigation with most crops on most soils, with little risk of developing a salinity problem. Some leaching is needed, but except in very impermeable soils this will occur with normal irrigation	< 2.5	< 0.2	< 200	< 60
C2	Medium salinity water. Can be used if moderate leaching is carried out. Plants of moderate salt tolerance can usually be grown without the need for special salinity control measures	2.5–7.5	0.2–0.5	200–500	60–200
C3	High salinity water. Cannot be used if drainage is restricted. Even if drainage is adequate, special management may be needed to control salinity, and plants with good salt tolerance are needed	7.5–22.5	0.5–1.5	500–1,500	200–600
C4	Very high salinity water. Unsuitable for irrigation, except under very special circumstances. Soils must be permeable, drainage adequate, and excess irrigation is needed to provide considerable leaching. Only very salt-tolerant crops should be grown	> 22.5	1.5–3.0	> 1,500	> 600

Table 2.16 USDA classification of irrigation water sodicity (Richards, 1954)

Sodium class	Description	SAR
S1	Low sodium water. Usable for irrigation on almost all soils with little danger of developing harmful levels of exchangeable sodium. Sodium-sensitive crops, for example, stone fruit, may accumulate injurious concentrations of sodium.	0–10
S2	Medium sodium water. Has an appreciable sodium hazard in fine-textured soils with high CEC, especially under low-leaching conditions, unless gypsum is added. May be used in coarse or organic soils with good permeability.	10–18
S3	High sodium water. May produce harmful levels of exchangeable sodium in most soils and will require good soil management, good drainage, high leaching and the addition of organic matter. Gypsiferous soils may not develop harmful levels of exchangeable sodium from such waters. Chemical amendments may be needed to replace exchangeable sodium, but may not be feasible if the water also has high salinity.	18–26
S4	Very high sodium water. Generally unsatisfactory for irrigation except under low or possibly medium salinity conditions, where the solution of calcium from the soil or the use of gypsum or other amendments may make it feasible.	> 26

Care also needs to be taken when analyzing groundwater samples. These tend to be lower in pH than surface waters due to their equilibration with much higher partial pressures of CO_2 (pCO_2). When exposed to atmospheric CO_2, the pH rises and precipitation of calcium compounds might occur. In such cases, the adjusted SAR needs to be used.

Water of high pH (> 8.5) has an excess of alkalinity over Ca and usually poses a substantial sodicity hazard. If pH is below 8.5 it could also have high alkalinity depending on the pCO_2. Very low salinity (EC_e around 0.1 dS m^{-1}) also makes waters more hazardous in terms of sodicity. Very high pH (> 9.0) adversely affects infiltration and limits Ca concentrations and high SAR. Water sodicity hazard also depends on its management, and greater infiltration problems are mostly associated with sprinkler irrigation than with other methods.

Figure 2.12 summarizes the various criteria for the evaluation of water quality, using the salinity and sodicity classes from Tables 2.15 and 2.16. It should be noted that, conventionally, the electrical conductivity of irrigation water is described in μS cm^{-1}. It is important to note that enough calcium may sometimes be dissolved from calcareous soils to decrease the Na hazard significantly, and this needs to be taken into account when using these waters, particularly those of class C1–S3 and C1–S4. For calcareous soils of high pH, or for non-calcareous soils, sodium status may be improved by adding gypsum to the water for waters in classes C1–S3, C1–S4 and C2–S4. For waters of C2–S3 and C3–S2, gypsum added to the soil is beneficial.

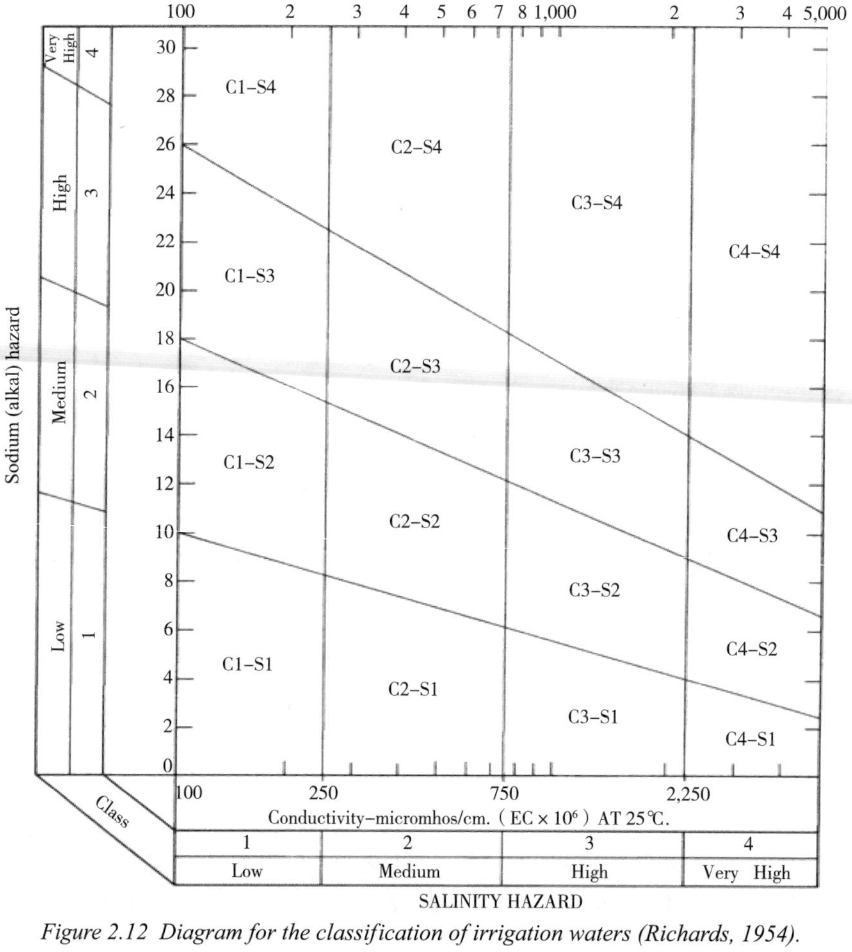

Figure 2.12 Diagram for the classification of irrigation waters (Richards, 1954).

Soils become sodic if they are irrigated with water of high "sodium hazard". In irrigation water, beneficial Ca^{2+} and Mg^{2+} can be precipitated out as insoluble compounds if the water has high CO_3^{2-} or HCO_3^- concentrations. The residual sodium carbonate (RSC) is calculated to determine irrigation water quality in terms of whether it has an excess of Ca^{2+} and Mg^{2+}, or of CO_3^{2-} or HCO_3^-. It is determined as:

$$RSC = (CO_3^{2-} + HCO_3^-) - (Ca^{2+} + Mg^{2+}) \tag{2.11}$$

where the items in parentheses are ion concentrations in mEq L^{-1}.

An RSC < 1.25 mEq L^{-1} means that the water is probably safe for irrigation. If it is between 1.25 and 2.5 mEq L^{-1}, then the water is marginal for irrigation, and if greater than 2.5 mEq L^{-1} then the water is unsuitable for irrigation.

The computer program WATSUIT can be used to predict the salinity, sodicity and toxic-solute concentration of the soil-water resulting from the use of particular irrigation water of given composition and at a specified leaching fraction. It can be also used to predict the effect of a particular salinity level on crop yield and of a given sodicity level on soil permeability. Like the other models, the program is available at http://ars.usda.gov/Services/docs.htm?docid = 8968.

ACKNOWLEDGEMENTS

We are grateful to Mr Chris Morgan and the Victoria Department of Primary Industries, Australia, for permission to use Figure 2.3 from the "Training manual on the use of the EM38 for soil salinity appraisal, Technical Report 181, Victorian Department of Primary Industries. © State of Victoria, Department of Primary Industries 1990".

REFERENCES

Ayers, R. S., & Westcot, D. W. (1976). *Water quality for agriculture*, FAO Irrigation and Drainage Paper 29, 1st edition, Food and Agriculture Organisation of the United Nations, Rome, Italy, 97.

Ayers, R. S., & Westcot D. W. (1985). *Water quality for agriculture*, FAO Irrigation and Drainage Paper 29, revision 1, Food and Agriculture Organisation of the United Nations, Rome, Italy, 174. Available at http://www.fao.org/docrep/003/T0234E/T0234E00.htm (April 7, 2008).

Barrett-Lennard, E. G. (2003). The interaction between waterlogging and salinity in higher plants: causes, consequences and implications. *Plant Soil*, *253*, 35-54.

Colorado River Basin Salinity Control Forum. (1993). *1993 review: water quality standards for salinity*, Colorado River System, Colorado River Basin Salinity Control Forum, Bountiful, Utah, 115.

Dregne, H. E., & Chou, N. T. (1992). Global desertification dimensions and costs, *Degradation and Restoration of Arid Lands*, In Dregne, H. E. (Eds.), Texas Tech. Univ. Press, Lubbock, Texas, USA, 249-282.

Dregne, H. E., et al. (1996). Soil salinity in China. *Desertification Control Bulletin*, *28*, 28-33.

Dudal, R., & Purnell, M. F. (1986). Land resources: salt-affected soils. Elsevier, *5*, 1-9.

El-Ashry, M. T., et al. (1985). Salinity pollution from irrigated agriculture. *J. Soil Wat. Cons*, *40*, 48-52.

FAO. (2007a). *The State of Food and Agriculture 2007*, Food and Agriculture Organisation of the United Nations, Rome, Italy, Available at ftp://ftp.fao.org/docrep/fao/010/a1200e/a1200e00.pdf.

FAO. (2007b). *Water at a Glance*, Food and Agriculture Organisation of the United Nations, Rome, Italy, Available at http://www.fao.org/nr/water/docs/wateataglance.pdf, (April 29, 2008).

FAO. (2007c). *Aquastat Database*, Food and Agriculture Organisation of the United Nations, Rome, Italy, Available at http://www.fao.org/nr/water/aquastat/dbase/index.stm, (April 30, 2008).

George, R. J., & Green, A. (2000). *Position paper on airborne geophysics for salinity and land management*, Airborne Geophysics Advisory Group, Sustainable Land and Water Resources Management Committee, Agriculture Western Australia, 32.

Ghassemi, F., et al. (1995). *Salinization of Land and Water Resources: Human Causes, Extent, Management and Case Studies*, CAB International, Wallingford, UK, 526.

Heynike, J. J. C. (1981). *The Economic Effects of the Mineral Content Present in the Vaal River Barrage on the Community of the PWVS Complex (A Desk Study)*, Water Research Commission, Pretoria, South Africa, 131.

Ijaz, K., & Davidson, A. P. (1997). *Baseline socio-economic survey - Joint Satiana Pilot Project*, International Waterlogging and Salinity Research Institute, Lahore, Pakistan.

James, C. V. (1989). Information China: The Comprehensive and Authoritative Reference Source of New China. *The Journal of Asian studies*, *48*(4), 822-824.

Joshi, P. K., et al. (1996). *Saline and waterlogged soils: Impact on agricultural economy and feasibility of reclamation*, Reclamation and management of waterlogged saline soils, National Seminar Proceedings, 5-8 April 1994, CSSRI, Karnal, India, 384-398.

Kielen, N. C. (1996). *Farmers' ability to cope with salinity and sodicity: farmers' perceptions, their strategies and practices for dealing with salinity and sodicity in their farming systems*, Research Report R6, International Irrigation Management Institute, Lahore, Pakistan, 78.

Landon, J. R. (1991). *Booker Tropical Soil Manual: a handbook for soil survey and agricultural land evaluation in the tropics and subtropics*, Booker Agriculture International Ltd., London, UK, 475.

Loveday, J. (1972). *Moisture content of soils for making saturation extracts and effect of grinding*, Division of Soils, Technical Paper No. 12a, Commonwealth Scientific and Industrial Research Organization, Canberra City, A.C.T. 2601, Australia.

Murray-Darling Basin Ministerial Council, (1989). *Draft Murray-Darling Basin Natural Resources Management Strategy*, Murray-Darling Basin Ministerial Council, Canberra, Australia, 19.

Norman, C. (1990). *Training Manual on the use of the EM38 for Soil Salinity Appraisal*, Tech. Rept 181, Victorian Department of Primary Industries, Australia.

Oldeman, L. R., et al. (1991). The extent of human-induced soil degradation, *World map of the status of human-induced soil degradation: an explanatory note*, In Oldeman, L. R., In Hakkeling, R. T. A., & Sombroek, W. G. (Eds.), International Soil Reference and Information Centre, Wageningen, Netherlands, 27-33.

Qureshi, R. H., & Barrett-Lennard, E. G. (1998). *Saline Agriculture for Irrigated Land in Pakistan: a Handbook*, ACIAR Monograph No 50, Australian Centre for International Agricultural Research, Canberra, Australia, 142.

Rafiq, M. (1975). *Saline, saline-alkali and waterlogged soils of the Indus Plains-their characteristics*, causes of formation and measures needed for reclamation, Proc. Intl. Conf. Waterlogging and Salinity, In Nawaz, M. N. (Eds.), Lahore, Pakistan, 308-321.

Rhoades, J. D., et al. (1992). *The use of saline waters for crop production*, *FAO Irrigation and Drainage Paper 48*, Food and Agriculture Organisation of the United Nations, Rome, Italy, 94. Available at ftp://ftp.fao.org/agl/aglw/docs/idp60.pdf (April 7, 2008).

Rhoades, J. D., et al. (1999). *Soil salinity assessment*, Methods and interpretation of electrical conductivity measurements, FAO Irrigation and Drainage Paper 57, Food and Agriculture Organisation of the United Nations, Rome, Italy, 160. Available at ftp://ftp.fao.org/agl/aglw/docs/idp57.pdf (April 7, 2008).

Richards, L. A. (1954). Diagnosis and improvement of saline and alkali soils, *USDA Handbook 60*, US Salinity Laboratory, Riverside, California, USA, 160.

Simmons, P., et al. (1991). *Management of irrigated water in the Murray-Darling Basin*, Discussion paper 91.6, Australian Bureau of Agricultural and Resource Economics, Canberra, Australia, 42.

Singh, N. T. (1998). Historic perspective, (Chapter 2), *Agricultural Salinity Management in India*, In Tyagi, N. T., & Minhas, P. S. (Eds.), Central Soil Salinity Research Institute, Karnal, India, 9-19.

Suarez, D. L. (1981). Relation between pHc and Sodium Adsorption Ratio (SAR) and an alternate method of estimating SAR of soil and drainage waters. *Soil Sci. Soc. Amer. J*, *45*, 469-475.

Szabolcs, I. (1989). *Salt-Affected Soils*, CRC Press, Boca Raton, Florida, USA, 274.

Tanji, K. K. (1990). Nature and extent of agricultural salinity (Chapter 1), *Agricultural Salinity Assessment and Management*, In Tanji, K. K. (Eds.), ACSE Manuals and reports on engineering practice No. 71, ASCE, New York, USA, 1-17.

US Salinity Laboratory. (2006). *The ESAP model*, Online software programmes available at http://www.ars.usda.gov/Services/docs.htm?docid=8918.

Vermeer, E. B. (1977). *Water Conservancy and Irrigation in China: Social, Economic and Agrotechnical Aspect*s, Leiden University Press, The Hague, Netherlands, 350.

WAPDA. (1988). *Water and Power Development Authority, Lahore, Pakistan, Personal Communication cited in Ghassemi et al.*, 1995.

Western Australia Legislative Assembly. (1991). *Select Committee into Land Conservation*, Final report, Legislative Assembly, Perth, WA, 171.

World Resources Institute (WRI). (2008). *World Resources Institute*, Online databases available at http://earthtrends.wri.org, (April 29, 2008).

CHAPTER 3.

EFFECTS OF SALINITY ON PLANTS

Philip Hollington

CAZS Natural Resources, School of Natural Sciences, Bangor University, Thoday Building, Deiniol Road, Bangor LL57 2UW, Wales, UK

1. INTRODUCTION

Salinity of the soil and/or irrigation water is one of the major abiotic stresses affecting crop production worldwide. This chapter looks in detail at the ways in which excess salts affect plant growth and survival, the physiological mechanisms and genetic control of salt tolerance, focusing on crop plant species as well as assessment of crop salinity tolerance.

From the salinity point of view, there is a fundamental division of plant species into two main classes: (i) the halophytes, which are adapted to grow in saline areas; and (ii) the glycophytes, which are not. Most cultivated plant (crop) species fall into the glycophyte category.

1.1 Halophytes

Halophytes (a term from the Greek for "salt plants") have evolved in salt-affected regions of the world. They generally show increased growth at low salt concentrations compared with no salt at all, but their growth is decreased at very high concentrations. Despite this, many grow at very high salinities indeed. The halophytes are widely distributed, and make up about one-third of angiosperm (flowering plant) families. Both monocotyledonous and dicotyledonous examples are found, and halophytes are particularly common in the Chenopodiaceae. Typical examples of halophytes include the river saltbush (*Atriplex amnicola*), which has a 10% growth increase at 5 dS m^{-1}, suffers a 50% decrease at 40 dS m^{-1}, but it is still alive at 75 dS m^{-1} (Qureshi and Barrett-

Lennard, 1998). Other species include quailbrush (*Atriplex lentiformis*), *Suaeda fruticosa*, and *Salicornia bigelovii*. Another typical example is *Kochia indica*, an indicator species which, if it is growing, shows that the land is almost certainly saline. Few, if any, agricultural crops can be regarded as halophytes.

1.2 Glycophytes

The term glycophyte is again from the Greek, this time for "sweet plants". Glycophytes are more common than halophytes, and supply the great majority of agriculturally important plants, although most cope poorly with salinity or sodicity. They are grouped into two main categories according to their tolerance to salinity. Salt-tolerant glycophytes can maintain growth at low salinity but show decreased growth at higher levels. A typical example is cotton (*Gossypium hirsutum*), in which the salinity level at which growth is reduced by 50% (EC_{50}) is 17 dS m^{-1}. Other species include barley (*Hordeum vulgare*, EC_{50} = 18 dS m^{-1}), sugarbeet (*Beta vulgaris*, 15 dS m^{-1}), and date palm (*Phoenix dactylifera*, EC_{50} = 18 dS m^{-1}). Salt-sensitive glycophytes are those in which growth is sensitive even to low salt concentrations. They include beans (*Phaseolus vulgaris*) and many other legumes, with a typical EC_{50} of around 3.5 dS m^{-1}. Other examples of salt-sensitive glycophytes include rice (*Oryza sativa*, EC_{50} = 7.2 dS m^{-1}), carrot (*Daucus carota* subsp. *sativus*, EC_{50} = 4.6 dS m^{-1}), grape fruit (*Citrus paradisi*, EC_{50} = 4.9 dS m^{-1}), and peach (*Prunus persica*, EC_{50}=4.1 dS m^{-1})

2. EFFECTS OF SALTS ON PLANTS

2.1 Plant response to salinity

Salinity adversely affects plants in two ways. First changes in the soil-plant-water relationship (a physical effect) will happen and specific chemical/biochemical effects due to the excess concentrations of certain ions follow. Although the salinity effects are usually plant-damaging, they may occasionally have beneficial effects. For example, slight salinity may increase yields of cotton, or improve tomato (*Solanum lycopersicum*) quality, and it has been shown to increase citrus tolerance to freezing.

2.1.1 The time sequence

Plant responses to salinity have been summarized by Munns (2002), from which this section has been drawn upon extensively. This author used plants grown either in sand or in solution culture to elucidate the effects. Salinity initially reduces root water potential leading to a reduced flow of water from the soil into the plant. The first consequence is a rapid but temporary drop in growth rate due to short-term changes in water relations, followed by gradual recovery to a new but reduced growth rate. Effects after this tend to be more permanent, and involve alterations to metabolism and the specific effects of particular ions.

The depression of external water potential by salinity, which occurs instantaneously with the increase in salinity [high osmotic pressure (OP)] narrows the gap between the external (soil) and internal (plant) water potentials, and as a result the leaves show an almost immediate reduction in growth rate. The reduction of growth rate in young leaves is so fast that it can only be due to changes in water relations (Passioura and Munns, 2000). Similar effects as those from NaCl can be induced using other osmotica, for example, KCl or mannitol (Yeo et al., 1991) showing that the changes are not "salt"-related. Similar effects have been induced in root growth, again using either NaCl or other osmotica (Frensch and Hsaio, 1994; 1995), thus the changes are not due to "salt" but to altered water relations. However, some indications of a "salt"-induced effect on root growth have been noted, as "salt"-induced Ca^{2+} deficiency is greater for genotypes with inherent low uptake of Ca^{2+} (Cramer et al., 1988; Colmer et al., 2006). After several minutes, growth recovers gradually to a new steady-state, although this is lower than before the application of salinity stress, particularly in the case of the roots. The concentration of the salt solution affects both the level of the new steady-state, and the time taken for this recovery.

After a few days, leaf and root growth have usually stabilized at their new, reduced, steady rate. This is usually lower for leaf growth than for roots, especially in dry soil (Hsiao and Xu, 2000; Munns and Sharp, 1993), again probably due to water stress rather than salts. The evidence for this is that Na^+ and Cl^- concentrations in the growing cells are always below toxic concentrations, both in the leaves (Hu and Schmidhalter, 1998) and in the roots (Jeschke, 1984; Jeschke et al., 1986). It is likely that the rapid expansion of the growing cells helps to maintain Na^+ and Cl^- at low levels.

After several days' exposure to high salinity, salt-specific effects may begin to occur in particularly salt-sensitive species. For example, some plants are unable to exclude NaCl when it is at high levels in the growing medium, and as a result show severe injury to the older leaves at this time. This is due to the accumulation of excessive levels of Na^+ and Cl^- in the transpiring leaves. When this exceeds the capacity of the cells to compartmentalise them in the vacuole, the ions rapidly build up, either in the cell walls where they dehydrate the cell, or in the cytoplasm where they inhibit enzyme activity.

If the ion concentration in the cell wall rises, the cell shrinks and the internal ion concentration rises. The efflux of some of these ions adds to those in the transpiration stream coming from the roots so that the ion concentration in the cell wall rises even faster and the cell dehydrates rapidly. If the site of increased concentration of Na^+ and Cl^-, once the vacuole is full,

distinguishable from deficiencies. These are particularly common in woody plants when they occur at even moderate concentrations (Bernstein, 1965). Excess Na^+ may cause several disorders affecting the Ca^{2+} transport and membrane function, which can occur very rapidly in some species and may sometimes be ameliorated by the addition of Ca^{2+} to the growing medium.

2.1.4 Trace elements

In salt-affected soils, deficiencies of a number of trace elements such as copper (Cu), iron (Fe), manganese (Mn) and zinc (Zn) are likely to occur. In saline soils, deficiencies of boron (B) are rarely a problem, although B toxicity may occur in some areas. Some trace elements may also accumulate in the tissues of plants growing in saline soils to levels that may be toxic to either animals or humans. Trace element levels in irrigation water also need to be monitored carefully to prevent adverse effects.

2.1.5 Summary of responses to salinity

Plant response to salinity occurs at different time scales, and the main effects of salinity on salt-tolerant plants are basically those due to soil water deficits, as shown in Table 3.1 below (Munns, 2002).

Table 3.1 Response of plants to salinity with time (after Munns, 2002)

Time scale	Water-stress (osmotic) effects	Salt-specific effects
Minutes	Immediate and rapid reduction in leaf and root growth, followed by a partial and rapid recovery	
Hours	Steady but reduced rate of leaf and root growth	
Days	Greater reduction in leaf growth than in root growth. Reduction in rate of leaf emergence	Visible signs of injury in the oldest leaves
Weeks	Smaller final leaf size. Possibly also reduced the number of lateral shoots	Death of the oldest leaves
Months	Effects on flowering time. Reduced seed production	Younger leaves dead, and possible death of the whole plant before maturity

Only the more sensitive plants are affected by salt-specific effects. Munns (1993) proposed a 2-phase salinity response model, which has been validated in wheat (Munns et al., 1995) and in maize (*Zea mays*) (Cramer et al., 1994; Fortmeier and Schubert, 1995). Similar effects were also noted in rice (Yeo et al., 1990). However, recent advances in the understanding of the osmotic effects of salinity have led to the model being refined, and a new version (Figure 3.1) was proposed by James et al. (2008).

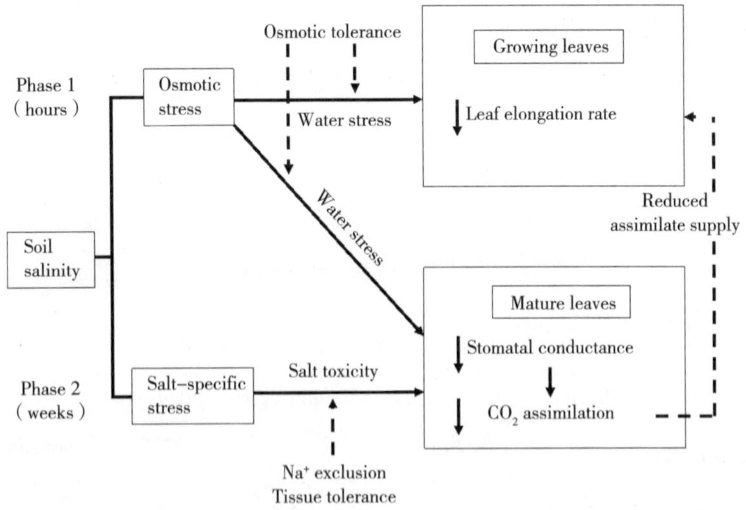

Figure 3.1 Revised 2-phase model for plant growth under salinity (James et al., 2008), developed from the initial model of Munns (1993). Reproduced by permission of CSIRO Publishing.

The initial plant response to salinity is water stress due to increased salt concentration (high osmotic pressure) in the root-zone (Phase 1), followed by salt toxicity resulting from the accumulation of salts (salt-specific stress) in the mature leaves (Phase 2).

The osmotic stress produces hormone signals that result in reduced leaf elongation rates (Termaat et al., 1985; Passioura and Munns, 2000), and also probably reduce the stomatal conductance. This results in reduced photosynthesis, which reduces the supply of carbohydrates to the growing leaves and so further reduces their growth. Salt accumulation in the older leaves eventually reaches toxic levels and so further inhibits photosynthesis, with reduced carbohydrate supply as the main factor limiting leaf and plant growth.

The various changes can be summarized as follows.

- Seconds or minutes after exposure to salt the cells lose water and shrink
- Within hours, they regain their original volume but with reduced elongation rates, and there are lower rates of leaf and root growth
- In days, there are changes in cell elongation and division leading to lower leaf appearance rate and smaller leaf size. Leaf growth is usually more affected than root growth, and where there is high salt uptake, the oldest leaf may begin to show injury
- In weeks, the lateral shoots are clearly inhibited, and in plants with high salt uptake many leaves may be dead, although rates of leaf production may not differ between genotypes
- In months, there are clear growth differences between plants with high and low salt uptakes, and there may be high leaf injury, reduced seed production and complete death in some cases at high salinity levels

2.2 Effect of sodicity on plants

Sodicity, the result of high concentrations of Na^+ ions relative to Ca^{2+} and Mg^{2+} in the soil cation exchange capacity or in the soil solution affects plant growth in two ways: alteration of the plant mineral nutrition due to Na^+ toxicity and disruption of the plant water balance due to its effects on the soil permeability. Crop tolerance to different levels of sodicity (ESP) is shown in Table 3.2.

Table 3.2 Crop tolerance to ESP (Bower, 1959)

ESP (%)	Type of crop affected	Growth responses under field conditions	Crop examples
2–10	Extremely sensitive	Sodium toxicity symptoms even at low ESP values	Deciduous fruit, nuts, avocado, cassava, citrus
10–20	Sensitive	Stunted growth at low values even though the physical condition of the soil may be good	Beans, sugarcane
20–40	Moderately tolerant	Stunted growth due to both nutritional factors and adverse soil conditions	Clover, oats, tall fescue, rice, dallis grass
40–60	Tolerant	Stunted growth is usually due to adverse soil physical conditions	Wheat, cotton, lucerne, barley, tomatoes, beets
>60	Most tolerant	Stunted growth is usually due to adverse soil physical condition	Crested fairway and tall wheatgrasses, Rhodes grass

2.3 Salinity tolerance physiological mechanisms

Salt tolerance is a complex, multigenic trait (Shannon, 1997; Flowers, 2004). Most crops are adversely affected by salinity, although in general, cereals are more tolerant than legumes (Reynolds et al., 2005), and many wild relatives of crop plants show greater tolerance than their domesticated descendents. Of the major cereals, bread wheat (*Triticum aestivum* L.) is generally regarded as moderately tolerant, durum wheat (*Triticum turgidum* subsp. *durum*) as somewhat less so, while rice (*Oryza sativa*) is regarded as susceptible (Francois and Maas, 1994).

Salt-tolerant and salt-sensitive plants are distinguished from one another by the inability of the latter to stop salts reaching toxic levels in transpiring leaves. The enzymes in salt-tolerant halophytes are no more tolerant to salt than are those in glycophytes (Greenway and Osmond, 1972; Flowers et al., 1977), and are inhibited by both Na^+ and Cl^- concentrations above about 100 mol m^{-3} (Munns, 2002). Tolerance mechanisms must, therefore, either minimize salt entry to the plant, or minimize salt concentration in the cytoplasm. Halophytes have both mechanisms. They are efficient salt excluders, and also compartmentalise into the vacuoles any salt that does enter. Some glycophytes are also good salt excluders, although unable to compartmentalise salt taken up as well as do the halophytes, but most are quite poor excluders, with the result that salt reaches toxic levels in the transpiring leaves.

In wheat, the key traits for salt tolerance were discussed by Colmer et al. (2005): one of the most important is a combination of low rates of Na^+ (high K^+/Na^+ discrimination) and Cl^- uptake and transport to the leaves (salt exclusion) (Munns et al., 2003). However, in rice, transpirational flow is highly correlated with Na^+ uptake and is an order of magnitude greater than in wheat (Garcia et al., 1997), reducing the importance of K^+/Na^+ discrimination in that crop.

2.3.1 Salt exclusion — low transport to leaves

Salt exclusion reduces the accumulation of salt in transpiring organs. Plants transpire between 30 and 70 times more water than is used for cell expansion, so any solutes that are not excluded by the roots would be between 30 and 70 times more concentrated in the plant than in the soil solution. An efficient filtering mechanism at the roots reduces this substantially. This was demonstrated in the early 1980s by Munns et al. (1983). Chloride concentrations in the xylem are between 0.2% and 5% of those in the external medium, which means that the roots filter out 95%–99.8% of the salt in the soil. If a plant transpires 50 times more water than it retains, but filters out 98% of the salt at the roots, then a shoot ion concentration greater than that of the soil will never be reached, and the plant could grow indefinitely in saline soil (Munns, 2002). Bread wheat is a good example of a plant that shows efficient Na^+ exclusion, and maintains a high K^+/Na^+ ratio in the shoots (Gorham et al., 1990). However, individual leaves cannot be protected indefinitely, and as the plant transpires, dissolved salts are deposited in the leaves as water passes out of the plant. These concentrations gradually increase, and eventually there will be much higher salt concentrations in older (i.e. those that have been transpiring for longer) than in younger leaves, which will eventually be high enough to kill the cells.

There are three main mechanisms to exclude salts from the leaves. The first is ion uptake selectivity at the roots. The cells that control this are not yet known, but it is believed that the initial Na^+ and Cl^- uptake could occur either at the epidermis, the exodermis or the endodermis. The second one is xylem loading, and Gorham et al. (1990) provided evidence for genetically-controlled loading of K^+ rather than Na^+ in the stele cells. The third mechanism is salt removal from the xylem in the upper roots, or the stem, petiole or leaf sheaths. In many species, Na^+ is retained in these areas, suggesting that K^+ is exchanged for Na^+ in the cells lining the transpiration stream.

In halophytes, salt exclusion is the most important regulator of the internal salt load, even in those species with salt glands or bladders. It is particularly important in perennial species with long-lived leaves, as these clearly need to regulate salt uptake for longer periods than annuals, whose leaves may be alive for only a few months. Some glycophytes can also control salt transport to the leaves fairly well, but only at low or moderate salinity levels. For example, barley excludes 95% of both Na^+ and Cl^- at the xylem level when grown in 100 mol m^{-3} NaCl (Munns, 1985).

There is now considerable evidence for the importance of these exclusion mechanisms. For example, see Greenway and Munns (1980) and specifically Laüchli (1984) for legumes, Storey and Walker (1999) for citrus species, and Jeschke (1984) for K^+/Na^+ selectivity in plants. There is a good understanding of the electrophysiology of the channels and transporters that regulate salt movement across the membranes as well as the genes that are involved in the process (Amtmann and Sanders, 1999; Hasegawa et al., 2000; Byrt et al., 2007).

A number of other features also help maintain low rates of salt accumulation in the leaves. For example, high shoot : root ratios and high intrinsic growth rates reduce the rate at which salt enters the transpiration stream and accumulates in the shoot (Pitman, 1984). The extent of the apoplastic pathway in the roots also influences salt movement across the root and into the xylem (Garcia et al., 1997). A third trait is the export of salts from the leaves via the phloem, although this is relatively minor compared with salt import via the transpiration stream. The evidence for this is that salt remains in the leaves for a long time after salts around the roots are removed (Munns, 2002). It appears that salt-tolerant species are better at excluding Na^+ and Cl^- from the phloem than are less tolerant ones, so ensuring that those ions do not reach the growing tissues. Some halophytes possess salt glands or bladders, which can help to export salts from the leaves (Flowers et al., 1977; 1986) and thus maintain salt balance over longer periods (Ball, 1988).

As a plant's capacity for salt exclusion is limited, above a particular threshold the exclusion mechanism breaks down, leading to high rates of transport of Na^+ and Cl^- to the shoots and leaf water deficits will develop. As a result, salt exclusion may not be entirely useful as a mechanism of salt tolerance. An example can be taken from Laüchli and Epstein (1990), who compared the uptake of Na^+ and Cl^- in sensitive and less sensitive genotypes of several crops: avocado (*Persea americana*) (Downton, 1978), grapevine (*Vitis* spp.) (Bernstein et al., 1969), maize (Schubert and Laüchli, 1986) and soybean (*Glycine max*) (Laüchli and Wieneke, 1979). In avocado, the less-sensitive cultivar excluded Na^+ from the shoot better than the sensitive one, although both took up similar amounts of Cl^-. In soybean, the less-sensitive cultivar excluded Na^+ and Cl^- better than the sensitive one. The tolerant grapevine was able to exclude Cl^- much more effectively than the sensitive one (measurements of Na^+ were not made), while in contrast the more tolerant maize cultivar showed poor exclusion of Na^+.

2.3.2 Intracellular ion compartmentation

Sequestration of Na^+ and Cl^- in the vacuole occurs in most species, as it is shown by the concentrations of these ions (> 200 mol m^{-3}), which would severely inhibit enzyme activity if the enzymes were exposed to them, that are found in still-functioning leaves. When sequestration occurs, K^+ and organic solutes accumulate in the cytoplasm to balance the osmotic pressure (OP) of the vacuole. The most frequent organic solutes involved are proline and glycinebetaine, although in some species other metabolites can reach high concentrations (Hasegawa et al., 2000). As well as accumulating in response to salinity, these compounds also accumulate under water stress, and are found at high concentrations in plants adapted to dry or saline soils. Their role is not entirely clear, as we know that they can accumulate to such high levels that they are not confined to the cytoplasmic compartments, and that their accumulation relates more to osmotic stress than to specific salt effects (Jones and Storey 1978).

Barley is a good example of a crop that uses intracellular ion compartmentation as its main salt tolerance mechanism. Both, Na^+ exclusion and K^+/Na^+ selectivity are lower than in bread wheat, but the Na^+ is effectively compartmentalized into the vacuole, so its toxicity is reduced (Greenway and Munns, 1980). Barley has an additional mechanism in that K^+ is compartmentalized into the mesophyll rather than the epidermis resulting in higher K^+/Na^+ ratios in the cytoplasm of the mesophyll, which helps to maintain photosynthesis.

Recently it has been realized that tissue tolerance to high Na^+ may be of much greater importance for breeding salinity tolerant varieties of wheat than had previously been thought (Genc et al., 2007). The work showed that, in a diverse collection of genotypes, there was no relationship between salt tolerance and Na^+ exclusion (as measured by K^+/Na^+ ratio), either from individual leaf blades or from the shoot as a whole, and Na^+ exclusion and Na^+ tissue tolerance were found to vary independently. They suggested that future efforts in selecting wheat for salt tolerance should concentrate both on Na^+ exclusion and tissue tolerance to high levels of Na^+.

2.3.3 The salinity tolerance energy costs to the plant

Tolerance mechanisms demand energy, and this is particularly important in the case of plants growing under saline waterlogged conditions. Three processes are involved, i.e. salt exclusion, intracellular ion compartmentation, and excretion through salt glands in certain species, and all have metabolic costs, although these are much smaller (around a whole order of magnitude) than those incurred by synthesising organic solutes for osmotic adjustment (Raven, 1985). This suggests that plants might grow faster in saline than in dry soils at the same water potential, although there is little evidence that this is in fact the case. Energy costs assume particular importance when considering combined salinity and waterlogging stress (see CHAPTER 3, section 3 EFFECTS OF COMBINED SALINITY AND WATERLOGGING ON PLANTS).

2.4 Genetic control of salt tolerance

Over the last few years a much clearer picture has emerged of the genes responsible for controlling salt tolerance. It is known that the *HKT* (high-affinity K^+ transporter) genes control sodium transport in higher plants (Munns and Tester, 2008), and these have been identified in wheat (Huang et al., 2006; Byrt et al., 2007), rice (Ren et al., 2005), barley (Haro et al., 2005) and *Arabidopsis* (Berthomieu et al., 2003). Huang et al. (2008) provided a detailed discussion of the role of these genes in salt tolerance and Na^+ transport in rice, wheat and barley.

Bread wheat has a low rate of Na^+ transport to the shoot and maintains a high K^+/Na^+ ratio in the leaves (Gorham et al., 1990). A trait for enhanced K^+/Na^+ discrimination at low salinity (Gorham et al., 1987), is conferred by the locus *Kna1* (Dvořák and Gorham, 1992), which is situated on the distal part of chromosome 4DL (Dubcovsky et al., 1996). It is likely that *Kna1* controls net xylem loading rather than net Na^+ uptake, as it does not produce differences in root Na^+ concentrations (Gorham et al., 1990). Quantitative trait loci (QTLs) for salt tolerance in bread wheat has been located on chromosome 5A (Semikhodskii et al., 1997), homoeologous to two yield QTL clusters on 5B and 5D in a different population grown under saline conditions (Quarrie et al., 2005).

In durum wheat, which in general has a poorer ability to exclude Na^+ (Gorham et al., 1990), a line with low leaf Na^+ concentrations and high K^+/Na^+ ratios has been identified (Munns et al., 2000). Munns et al. (2003) identified two genes, *Nax1* and *Nax2*, as being responsible for this trait. *Nax1* reduces the transport of Na^+ from the roots to the shoots and causes retention of Na^+ in the leaf blades (James et al., 2006). It has been mapped to chromosome 2AL (Lindsay et al., 2004), and a candidate gene identified as a sodium transporter *HKT7-A2* mapped to the same region (Huang et al., 2006). This is consistent with *Nax1*'s physiological role in reducing Na^+ in the leaf blades by retaining the ions in the leaf sheaths, and was postulated as controlling Na^+ unloading from the xylem in the roots and sheaths.

Nax2 also confers a lower rate of Na^+ transport from the roots to the shoots, but has a higher rate of K^+ transport than *Nax1*, leading to enhanced leaf K^+/Na^+ discrimination (James et al., 2006). It was suggested that the gene only functions in the roots, as lines containing *Nax2* did not retain Na^+ in the bases of the leaves (James et al., 2006). It has been mapped to the distal region of chromosome 5AL (Byrt et al., 2007), which coincides with the locus for a putative Na^+ transporter *HKT1;5-A* (*HKT8*). The *Nax2* region on 5AL is homoeologous to the region on 4DL that carries *KNa1*, and it is therefore highly probable that both *Nax2* and *Kna1* are strongly associated with *HKT1;5* genes. In consequence, Byrt et al. (2007) proposed *HKT1;5-D* as the candidate gene for this locus. Davenport et al. (2005) suggested that the two genes *Nax1* and *Nax2* interacted via net Na^+ xylem loading and net leaf sheath sequestration to control Na^+ exclusion.

In the *Thinopyrum* species, related to wheat, salt tolerance is associated with shoot Na^+ and Cl^- exclusion, particularly at higher NaCl concentrations (Gorham, 1994). In *Th. elongatum*, Dvořák et al. (1988) and Omielan et al. (1991) reported major effects of chromosomes 3, 4 and 7 on ion exclusion. Koebner et al. (1996) proved the involvement of chromosomes of homoeologous group 5 in salt tolerance in wheat. These effects are closely linked to genes controlling vernalization (group 5) and photoperiod (group 2) response (Taeb et al., 1992; Martin et al., 1993). In *Th. bessarabicum,* group 2 chromosomes carry genes conferring susceptibility to salt, and chromosome 5Eb has a major dominant gene or genes for salt tolerance (Forster et al., 1988; Mahmood and Quarrie, 1993). Zhong and Dvořák (1995a; 1995b) suggested that *Triticeae* species generally shared common mechanisms of tolerance to sudden salinity stress.

In rice, Flowers et al. (2000) identified AFLP markers for ion transport and selectivity. In the tolerant variety Pokkali, a major gene, possibly *SalT* (Causse et al., 1994), has been mapped (IRRI 1998), as were QTLs governing high K^+ and low Na^+ absorption, and high K^+/Na^+ ratio (Gregorio et al., 2002). A common major QTL was found on chromosome 1 for three traits associated with salt tolerance (IRRI, 1998). One QTL for salt tolerance on rice chromosome 1 has been fine mapped and the gene, *SKC1*, has been identified as a sodium transporter (Ren et al., 2005). It is not yet clear whether it is responsible for other QTLs on chromosome 1, or if there is a cluster of relevant genes in this region. Other markers in rice have been identified for traits associated with productivity in saline environments, although many are associated with tolerance to submergence, and micronutrient deficiency or toxicity (Gregorio et al., 2002). Koyama et al. (2001), Lin et al. (2004) and Takehisa et al. (2004) have also mapped other QTLs for rice salt tolerance.

Overall, relatively few QTLs have been identified for salt tolerance in some crop species. This could either be because the traits are actually determined by a limited number of loci or that the genes associated with the traits were clustered on particular chromosomes (Flowers, 2004).

2.5 Response to salinity during plant development

The sensitivity of plants to salinity changes according to environmental conditions and over the course of their development (Shannon, 1997) with the critical stages being germination, vegetative growth and reproductive growth. The vegetative growth of some halophytes is stimulated by salinity but they are not salt-tolerant at germination. While vegetative growth in glycophytes is generally sensitive to salinity, some crops are more tolerant during reproductive development, although the picture is complex.

Some salt-sensitive species, for example, maize (Maas et al., 1983) and tomato (Kurth et al., 1986; Foolad and Lin, 1997), may germinate well at high salinity. In triticale (× *Triticosecale*), tolerance varies between germination and emergence according to the genotype, and may be improved by calcium (Norlyn and Epstein, 1984). Both cotton and sugar beet are more salt-tolerant crops than triticale during their development, but they are more sensitive than the later at germination (Kent and Lauchli, 1985; Beatty and Ehlig, 1973). Cotton yield may be only slightly reduced by salt, despite large decreases in vegetative growth (Rains et al., 1987), possibly as a result of shifts in the hormonal balance. Grain yield in rice is more depressed by salt than its vegetative growth, and germination is relatively tolerant to salt (Khatun and Flowers, 1995). The developmental shifts in relative salt tolerance response vary with genotype. For example, in barley the cultivar Briggs that is more sensitive to early salinity was more tolerant at later stages while another cultivar showed opposite response to salinity (Lynch et al., 1982).

Several cereals, for example, wheat, sorghum and oats, develop and mature more rapidly under saline conditions, but others, for example, barley and rye, are unaffected (Shannon et al., 1994). Grieve et al. (1994) found that the phyllochron in spring wheat increased with salinity. Salinity decreased the number of tillers of spring and durum wheat (Maas and Grieve, 1990), and this was the main cause of yield reductions in both field and greenhouse-grown plants (Francois and Maas, 1994).

3. EFFECTS OF COMBINED SALINITY AND WATERLOGGING ON PLANTS

3.1 Extent of the problem

Waterlogging is commonly associated with secondary salinity, and it is reported as a major problem in some areas of Northwest India and Pakistan, for example (CSSRI, 1997). It is due to the presence of shallow water tables as a result of seepage from canals, monsoonal flooding or over-irrigation and/or poor physical conditions (low infiltration) created by the soil sodicity (Ghassemi et al., 1995). The problem is exacerbated in rice-wheat systems growing areas such as those of Northwest India and Pakistan due to the subsoil compaction that is done to produce an environment for flooded rice (Samad et al., 2001). The combination of waterlogging and salinity is far worse than the presence of either stress on its own. Barrett-Lennard (1986) suggested that a substantial proportion of salt-affected land worldwide was subject to occasional, intermittent or prolonged waterlogging.

3.2 Effects on plant ion uptake

When a soil is waterlogged, oxygen availability is greatly reduced. The diffusion rate of oxygen is about 10,000 times lower through water than it is through air, its solubility in water is very low, also bacteria and roots use up much of the oxygen that is in soil. The soil is said to have become anaerobic or hypoxic. In addition to this, carbon dioxide and various other products of root and bacterial anaerobic metabolism accumulate, leading to the development of reducing conditions in the soil. As a result, energy production from the breakdown of carbohydrate (respiration) is very much lower than is possible under aerobic (oxygenated) conditions.

As noted in CHAPTER 3, section 2.3.1 Salt exclusion–low transport to leaves, plants filter out salts at the root surface to a greater or lesser extent. However, this filtering process is very expensive in terms of energy, and anything that decreases its efficiency has a severe effect on plant growth and survival. Barrett-Lennard (1986) calculated that Na^+ exclusion from the roots required about 2.5% of the total energy available to a root in drained soil, and it is likely that Cl^- exclusion would need about the same. In aerobic soils, the oxidation process of the glucose in the roots produces up to 6 mol each of H_2O and CO_2 and up to 38 mol of ATP—adenosine triphosphate, the molecule that powers cell metabolism. However, in anaerobic waterlogged conditions, the process produces 2 mol each of ethanol and CO_2, and only 2 mol of ATP, a reduction of about 95% in the ATP production and hence in the fuel available for the filtration mechanism.

As a result, a major impact of waterlogging under saline conditions is to prevent the roots from screening out salts, so that there are large increases in salt uptake and shoot salt concentrations in the plant. The comprehensive review of Barrett-Lennard (2003) noted the following.

- Waterlogging interacted with salinity to increase shoot concentrations of both Na^+ and Cl^-. This was initially due to increases in the net transport of these ions to the shoots, and later due to reduced shoot growth and impairment of root functions
- This interaction led to increased Na^+ and Cl^- concentrations in the shoots, which had severe effects on plant growth and survival
- The interaction also had profound implications for the management of salt-affected areas and for breeding crops or plants for revegetation of these lands

3.3 Effects on plant growth

Previous exposure to waterlogging can improve a plant's ability to cope with saline waterlogging. This may be due to stimulation of the aerenchyma (tissue with large pores) formation in the roots (Kriedemann and Sands, 1984). These pores allow oxygen to diffuse through the roots, and so the plant is able to maintain energy levels.

Barrett-Lennard et al. (1999) concluded from laboratory experiments with wheat that the sequential effects of combined salinity and hypoxia on growth and senescence were probably due to adverse ion relations in the oldest leaves. These results were in agreement with those of a number of other authors (John et al., 1977; Barrett-Lennard, 1986; Marcar, 1993).

The sequence of events occurring after the onset of saline hypoxia in the root medium was as follows (assuming a solution with a NaCl concentration of 30 mol m^{-3} or greater): After one week, roots ceased to grow, and increased uptake of Na^+ and Cl^-, combined with decreased K^+ uptake, led to increased Na^+ and Cl^- concentrations in the oldest leaves and decreased K^+ concentrations in the youngest leaves. In a further week, the older leaves of the main tiller began to senesce, and after another week this process was at the same rate, or faster, than new leaves were being produced. In one month, or seven weeks after the onset of the stress, the plants would be more or less dead, and in any case unable to produce grain.

3.4 A potential screening technique for tolerance to salinity and waterlogging

The severe effects of combined salinity and waterlogging may be a reason for the lack of success in breeding successful crop varieties for saline lands. Breeding in the past has simply been for salinity and has not taken account of waterlogging as well. Barrett-Lennard et al. (1999) proposed a novel screening technique to overcome this by simultaneously testing for salinity and waterlogging tolerance. This would be done through a simple comparison of the rates of production and senescence of leaves on the main culm. The proposed method was tested in Faisalabad, Pakistan using 25 wheat genotypes (Hollington et al., 1999). Rates of leaf emergence and senescence were recorded over a 30 days period. The emergence and senescence rates for each genotype were plotted, and the point at which the two slopes crossed over was noted. The hypothesis was that genotypes with a long time to crossover, i.e. those in which the rate of emergence was much greater than that of senescence under saline hypoxia, would be tolerant to the combined effect of the two stresses, and that the technique will provide an easy and rapid way to identify these genotypes. Unfortunately, the relationship between rates of leaf emergence and senescence was not a good indicator of tolerance to salinity and waterlogging in terms of shoot weight, although there was some correlation with root weight. Although the method has potential, it still needs further refinement and more testing before it can be recommended.

4. EFFECTS OF SPRINKLER IRRIGATION WITH SALINE WATER

Everything so far in this chapter has referred to the effects of soil salinity or salinity in irrigation water, where the water is applied to the soil surface and is taken up through the roots. This type of flood irrigation is still the most common in many parts of the world. However, in more modern irrigated systems for high-value crops, some form of sprinkler application is often used. This has major implications in areas where the waters contain salts, as crop tolerance to saline water applied through sprinklers (to the aerial plant organs) is often very different from that when saline water is applied through flood irrigation. The problem is that a crop species/ variety that takes up or excludes Na^+ or Cl^- efficiently through the roots may not show the same efficiency through the leaves. This has long been known to be the case for fruit trees and other woody plants, as shown in Table 3.3, but it was not thought to be of concern in arable crops. However, Aragüés et al. (1994) noted that barley took up Cl^- more readily through the roots than through the leaves, and that maize in particular and barley also took up more Na^+ through the leaves than through the roots (Benes et al., 1996).

Table 3.3 Relative susceptibility of crops to foliar injury from sprinkler irrigation with saline water (Maas, 1990 cited in Rhoades et al., 1992)

Na^+ or Cl^- concentration causing foliar injury (mol m^{-3})			
<5	5–10	10–20	> 20
Almond	Grape	Lucerne	Cauliflower
Apricot	Pepper	Barley	Cotton
Citrus	Potato	Maize	Sugar beet
Plum	Tomato	Cucumber	Sunflower
		Safflower	
		Sesame	
		Sorghum	

5. ASSESSMENT METHODS OF CROP SALINITY TOLERANCE

There are three methods commonly used to assess the salt tolerance of a crop. These are done in terms of: (i) its survival; (ii) its absolute yield in saline conditions; or (iii) its relative yield compared with non-saline controls. Each of these methods has its use, but the relative yield is the most common assessment method for agricultural crops.

5.1 Survival

The survival of a crop is the best method ecologically. Although survival is often at the expense of growth and yield, it is pretty much useless from an agricultural point of view as it gives no measure of the agronomic usefulness of the plant.

5.2 Absolute yield in saline conditions

The absolute yield in saline conditions, i.e. the yield in terms of tonnes or kilograms per hectare, is the best when tolerance is looked at from an economic point of view. However, a major drawback is that it does not allow comparisons to be made between different crops, which may have very different growth habits and produce. A simple, if extreme, example will make this clear.

Both sugar cane (*Saccharum officinarum*) and wheat can be grown on saline land. In normal conditions, a crop of sugar cane should yield between 70 and 100 t ha^{-1} under rainfed conditions, and 110–150 t ha^{-1} under irrigation. Similarly, bread wheat could produce around 1.3–2.0 t ha^{-1} under rainfed, and 3.0–5.0 t ha^{-1} under irrigated conditions. There is, thus, an enormous difference between the yield potential of the two crops. However, at an EC_e of 10 dS m^{-1}, irrigated sugarcane might yield 50–60 t ha^{-1}, and wheat 2.50–3.75 t ha^{-1}. Even though the yield of sugarcane is higher than that of wheat, we cannot say it is a more tolerant crop, as its yield has been reduced much more than that of wheat. In economic terms, it might be a more worthwhile crop, but it is NOT more tolerant.

5.3 Relative yield compared with non-saline controls

The advantage of using relative yield is that crops can be compared where their yields are different. Relative yield can simply be defined as the yield under saline conditions as a proportion (or percentage) of the yield under non-saline conditions. It makes no difference whether we take the yield under non-saline conditions as 100% or as 1, so we can say the relative yield is, for example, either 60% or 0.6. The use of relative comparisons is not confined to the grain yield. We can use it for any part of the plant in which we are interested, for example, number of grains, height, total biomass, or oil yield. We are also not confined to using it for salinity. The same concept can be used under any other stress condition, for example, to assess drought tolerance or response to heat stress.

5.4 Crop salinity response models

Earlier work on assessing relative yields used the "bent stick" model of Maas and Hoffmann (1977). This method described the crop yield response as a function of the average root-zone salinity in terms of the threshold (the salinity level below which there was no effect on growth or yield) and the slope (the reduction per unit increase salinity in the yield or other trait of interest). Using this method, workers at the US Salinity Laboratory analyzed experimentally derived crop salt tolerance data and produced tables of crop salt tolerance, grouping crops as tolerant, medium tolerant, medium susceptible and susceptible. Examples of such tables include Maas (1984; 1990), and Maas and Grattan (1999), which have been used by workers all over the world. These tables also have a common problem as many of the varieties and genotypes used are old, are often not specified, and are not adapted to areas such as those in India and Pakistan where salt-tolerant crops are required. This needs to be borne in mind by anyone using them.

The crop salinity response function can be described in terms of a simple equation:

$$Y_r = 100 - b\,(EC_e - a) \tag{3.1}$$

where a is the salinity threshold in dS m^{-1}, b is the slope expressed in percent per dS m^{-1}, and EC_e is the mean conductivity of the saturated paste from the root-zone. An example of the model is shown in Figure 3.2 below, while Figure 3.3 shows how this model can be used to classify crop salinity tolerance, based upon their calculated threshold and slope values.

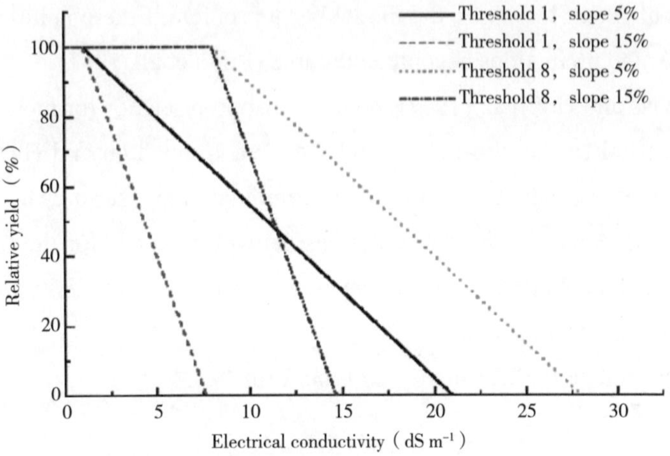

Figure 3.2 Example of the "bent stick" model of Maas and Hoffman (1977).

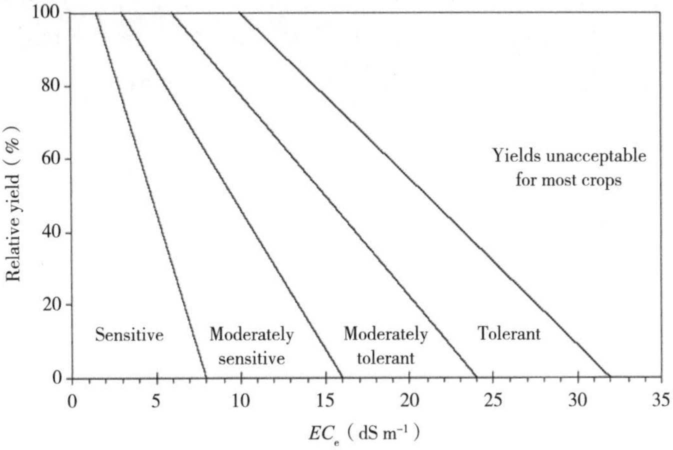

Figure 3.3 Division for classifying crop tolerance to salinity (Tanji and Kielen, 2002).

Figure 3.2 shows the response of four hypothetical crop genotypes to salinity. Genotype 1 has a threshold salinity level of 1 dS m^{-1}, and a slope of 5%, i.e. for every dS m^{-1} increment in conductivity, the relative yield declines by 5%. It is somewhat more tolerant than genotype 2, which has the same threshold of 1 dS m^{-1} but a slope of 15% and is clearly the least tolerant of the four genotypes. Similarly, genotype 3, with a threshold of 8 dS m^{-1} and a slope of 5% is more tolerant than genotype 4, which has a threshold of 8 dS m^{-1} and a slope of 15%. Genotype 3 is the most tolerant of all. It is debatable which of genotypes 1 and 4 is the most tolerant. It would depend upon the situation in the field. Under low salinity, then genotype 4 would be preferred, but at higher levels then genotype 1 would do better.

The Maas and Hoffman (1977) model is of value to give general tolerance data and to enable managers and farmers to make informed decisions. For example, it is important to know roughly the salinity level at which the yield of a certain crop or variety will start to decrease, and at what rate this will occur. However, the model has a problem if there is little data to the left of the threshold value, and predictions become unreliable (Royo et al., 1991).

Where the Maas and Hoffman (1977) model is inappropriate, or for more precise modelling of the crop response, then the non-linear model of van Genuchten and Hoffman (1984) may be preferred. This model produces a continuous sigmoidal response to changes in salinity. The computer programs to carry out these calculations, as well as those for the other models, were described by van Genuchten (1983). They are available from the US Salinity Laboratory at http://ars.usda.gov/Services/docs.htm?docid = 8957.

The most usual equation of the model is of the form:

$$Y_r = Y_m / [1 + (EC_e / EC_{50})^P] \tag{3.2}$$

where Y_r is the relative yield for a given EC_e, Y_m is the maximum yield that can be obtained under non-saline conditions, EC_{50} is the EC_e that reduces yield by 50% and P is a constant describing the shape of the curve. The values of EC_{50} and P are calculated by the model, and can also be calculated using non-linear regression in statistical software such as Genstat or SAS.

Figure 3.4 shows an example of the van Genuchten and Hoffmann model for the response of four hypothetical genotypes. Genotypes 1 and 2 have $EC_{50} = 8$ dS m^{-1}, and for genotypes 3 and 4 $EC_{50} = 16$ dS m^{-1}, i.e. the salinity at which yield is reduced by 50% is twice as high for genotypes 3 and 4 as for genotypes 1 and 2. The values of P, the coefficient that describes the shape of the curve, are $P = 2$ for genotypes 1 and 3, and $P = 4$ for genotypes 2 and 4. This time, genotypes 1 and 2 are clearly less tolerant than genotypes 3 and 4, but the distinction within the tolerance groups is less clear.

Figure 3.5 shows the data from an actual field experiment in Spain (Philip Hollington unpublished data). The data were obtained using the triple-line-sprinkler system described in

CHAPTER 3, Section 6 EXPERMENTATION FOR ASSESSMENT OF CROP SALINITT TOLERANCE, and was part of a trial to test the salt tolerance of wheat genotypes, including a novel addition line (Forster et al., 1988) incorporating an extra chromosome from *Thinopyrum bessarabicum* in Chinese Spring wheat (Chinese Spring 5Eb). The other genotypes were salt-tolerant Indian varieties KRL 1–4 and Kharchia 65, developed at the CSSRI, Karnal, Bahawalpur, a non-tolerant Pakistani variety, LU26S, a salt-tolerant selection made by Qureshi and Barrett-l Lennard (1998) in Faisalabad from the non-tolerant Pakistani variety LU26, and KTDH 19. This was a doubled haploid line developed by Quarrie and Mahmood (1993) at the John Innes Centre, UK from a cross between the tolerant Kharchia 65 and the highly-efficient Na^+-excluding line TW 161. Figure 3.5 also shows that the model can be used with absolute data, rather than just with relative yields.

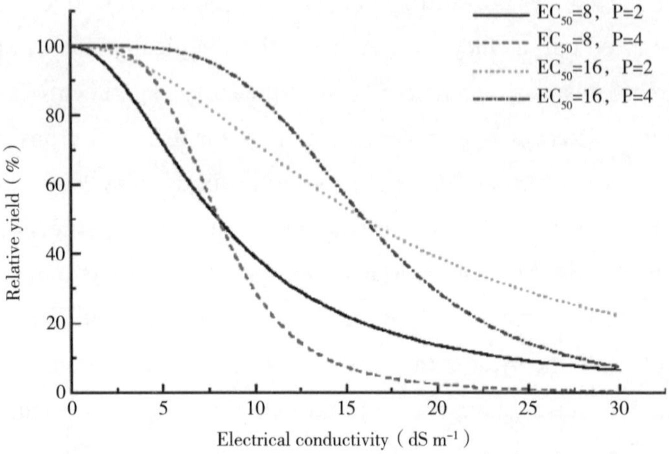

Figure 3.4 Example of the non-linear model of van Genuchten and Hoffman (1984).

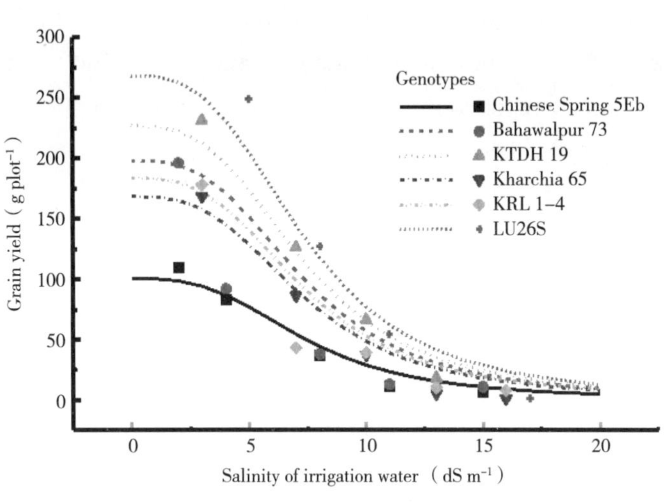

Figure 3.5 Response of wheat genotypes and hybrids to irrigation water salinity under controlled conditions in Spain (Philip Hollington unpublished data).

Figure 3.5 shows that, unexpectedly, on this occasion there was no yield advantage due to the additional chromosome in Chinese Spring. LU26S had the highest yield potential, but yield differences between genotypes became smaller as salinity increased.

6. EXPERIMENTATION FOR ASSESSMENT OF CROP SALINITY TOLERANCE

Field assessment of salinity tolerance is extremely difficult due to the high variability of salinity, both spatially and temporally. However, controlled-environment trials (for example those conducted in hydroponics, flood benches, sand culture, etc. in the greenhouse) do not provide a realistic prediction of the crop behaviour under field conditions. This is because of many differences in the stresses prevailing in the two situations. Even so, controlled environment trials should not be ruled out, as they are useful for physiological studies, in particular attempts to understand the mechanisms of tolerance. Most of the early experiments carried out at the US Salinity Laboratory to determine salt tolerance classes for different crops were done in such conditions, and they have proved useful, in general terms, for decades.

Whatever system is used, it is imperative to ensure that the experiment is testing tolerance to salinity, rather than to osmotic shock. Many published papers have applied the salinity all at once, rather than gradually increasing it, and do not include the addition of Ca^{2+} to the growth medium. As a result, what they test is the ability of the plant to withstand osmotic shock rather than salinity. Great scepticism should be used on data that show the results of trials in which salinity increases at a greater rate than 50 mol m^{-3} d^{-1}, for example. Other common failings in experimental techniques include a failure to conduct experiments under conditions where plants have limited or no transpiration. This is a common failing of work on cell culture and transgenics (Flowers, 2004). Assessments need to be made of plant growth and vigour, and ideally, for agricultural crops, results are reported in terms of the yield obtained in the field, as in the end this is the only thing that matters.

6.1 Field experiments

The overall objective is to minimize the experimental variation as much as possible in order to be able to distinguish differences between treatments (i.e. minimize the ratio of experimental to genotypic or treatment error). The following method was developed for testing salinity tolerance under natural field conditions in Spain (Ramon Aragüés, personal communication).

- Locate a plot with visual symptoms of salinity (for example, white patches in some areas, differences in surface texture) and, preferably with a regular gradient in crop yield decrements

• Construct a detailed salinity map of the plot. The only way to do this is through EC_a (apparent soil electrical conductivity) readings of the EM sensor on a regular (ideally 2 m × 2 m) grid. Delineate the EC_a isolines, preferably using geostatistical techniques (for example, using computer programs such as Surfer or ESAP)

• Sow the crop or cultivars along the direction of the maximum salinity gradient. Sow in bands of 4 or 5 rows, depending upon between-row spacing, in rows of at least 25 m

• Apply conventional cultural and agronomic practices. Regular practice of farmers is to give pre-sowing or post-sowing irrigations to leach the salts from the seedbed

• Take regular EC_a readings (for example every 15 days and/or after each irrigation) on the previously-used grid to ascertain the temporal variation of soil salinity

• Carry out several (e.g. 3) calibrations of the instrument during the growing season. Ensure the number of data points each time is greater than 10. Measure the EC_e of the soil in the crop root-zone to transform EC_a to EC_e using $EC_e = a + b \times EC_a$. From the average calibration, convert the EC_a maps to EC_e maps

• Harvest grain yield (or other target crop variable) in selected small areas by hand, preferably where the crop appears to be uniform (0.5 – 1.0 m^2 area). In an experiment in Spain, an area of 4 m^2 was harvested by machine with poor results. Harvest or sample over the maximum number of possible points (ideally 20 per genotype or other treatment) from areas of low to high yields

• Plot the yield – EC_e observations. These EC_e values are the time-weighted average data obtained from the EC_e maps delineated in point 4 above. Decide the salt response model that best fits these observations. If necessary, carefully delete "extraneous" (outliers) points. Estimate the salinity tolerance parameters (e.g. the threshold EC_e and the slope of the line, or the EC_{50} and values of p and Y_{max}

The author has used a modified system for field work in India and Pakistan as follows: A field with visible variability in salinity levels is selected and an EM38 survey and calibrations are carried out as before. If ESAP is not available, then sample positions can be selected using other methods, ensuring they are representative of the range of salinity in the field, and the plots can be made using any graphing package.

From the "contour" map of field salinity, areas of homogeneous salinity should be identified, and between eight and nine blocks selected in these. Relatively long but narrow plots should then be sown, with five or six rows and length 4 m or 5 m. After every irrigation additional sampling with the EM38 should be carried out at two or three points within each of these. At harvest, the yield and yield components should be assessed on two 1-m row lengths in the central three rows, again taking a salinity reading from the sampling point. The use of eight or nine blocks (replications) has been found to be the minimum necessary to ensure a reasonable ratio of treatment error to experimental variation. If necessary, precision can be increased further by adjusting the data statistically, using nearest neighbor analysis or covariance on adjacent plots.

6.2 Controlled field trials

Controlled field trials are a useful half-way house between the high variability of the natural field and the artificially controlled-environment of greenhouse experiments. Examples include triple-line sprinkler systems and controlled drip irrigation systems, which were developed to allow a more accurate control of field salinity without the drawbacks of the greenhouse.

The triple line source sprinkle (TLS) was developed and tested in Zaragoza, Spain (Aragüés et al., 1992; Royo and Aragüés, 1993). It consists of two outer laterals supplying fresh water (EC_e < 2 dS m^{-1}), and a third lateral midway between them to supply saline water (EC_e = 20 dS m^{-1}). The lateral spacing is equal to the sprinkler's wetted radius, so the layout duplicates a salinity gradient on each side of the centre line, and maintains a uniform supply of fresh water. The system provides 10 salinity treatments perpendicular to the laterals. The TLS was used for over 10 years, but over time a number of problems became apparent. The first was excess ion absorption through the leaves. Some crop varieties show different patterns of uptake by the roots compared to the leaves (e.g. Gorham et al., 1994 reported for barley), and ion absorption through the leaves could distort the assessment of tolerance. As a result, the TLS is not a realistic simulation of the situation in flood irrigated areas. In addition, the TLS could only be used under very light wind conditions (wind speeds < 2 m s^{-1}), and was comparatively labour-intensive to operate.

Despite altering the operating protocols to include pre- and post-irrigation washings of the leaves with fresh water, it was decided to develop a better system. This led to the first controlled drip irrigation system (Isla et al., 1997). This system consisted of three 3,000-l PVC tanks, a pumping system, three 75-mm diameter polyethylene lines (one for each saline treatment), and 25-mm diameter laterals with emitters located 0.2 m apart. Plot size was 1.45 m × 1.25 m, allowing for six rows of plants. Within each plot there were three irrigation laterals located 0.42 m apart, and 23 evenly-spaced emitters delivering a flow of 12 L h^{-1} to provide uniform and complete wetting of the soil. The system was used for salinity tolerance testing in the 1990s, but it was found that frequent irrigations led to shallow rooting of the crops. A new drip-injection irrigation system (DIS) was developed (Aragüés et al., 1999; Royo et al., 2000). This system consists of a parallel pump system with a centrifugal pump for fresh water and an injection pump for saline water, combined with a conventional drip irrigation system. The injection pump had a fixed discharge, while the discharge of the centrifugal pump was controlled by varying the number of emitters in each sector. The number of emitters installed in that sector therefore, controlled the salinity developed in each sector. The DIS was validated and found to be a low cost (at 1998 values about $5,000 automated and $3,000 manual) and effective system for evaluating salt tolerance under field conditions.

6.3 Implications of the response models on screening crop genotypes for salt tolerance

The response models described above provide a number of parameters, which have been proposed for use in screening crop genotypes for salt tolerance. The value of these will depend upon the particular situation being screened for. Ramon Aragüés and his group in Zaragoza, Spain, working mainly on barley under both controlled and natural field conditions, have carried out a great deal of work on this. Under controlled conditions using the TLS, both the linear and the non-linear models proved adequate for describing the response of a large number of barley genotypes to salinity (Royo and Aragüés, 1999). There was no correlation between EC_{50} and Y_m, suggesting that the most productive barley genotypes were not necessarily the least tolerant to salinity, in contrast to data from Shannon (1997) and Pasternak and de Malach (1994). In fact, the most productive genotypes under both moderate (6 dS m^{-1}) and high (12 dS m^{-1}) salinity were those with the highest yield potential, Y_m. It was also concluded that the salinity threshold, EC_t, was not an efficient parameter for screening. EC_t was not correlated with yields at 6 dS m^{-1}, and was negatively correlated with the yield at 12 dS m^{-1}, so selecting for high EC_t would select genotypes with low yields under high salinity conditions. In breeding barley under natural field conditions, Isla et al. (2003) concluded that Y_m should be used as the selection criterion on moderately saline soils (average EC_e 5–7 dS m^{-1}), but on soils of high salinity (EC_e around 15 dS m^{-1}) then selection should be based upon a combination index $EC_{50}\times(Y_m \times 10^{-3})$.

ACKNOWLEDGEMENTS

Figure 3.1 is reproduced with permission from *Functional Plant Biology* 35: 111–123 (R. A. James et al.), Copyright CSIRO 2008, published by CSIRO Publ., Melbourne, Australia. The full text of the article can be accessed via either subscriptions or pay-per-view services at http://www.publish.csiro.au/nid/103/issue/4078.htm.

REFERENCES

Amtmann, A., & Sanders, D. (1999). Mechanisms of Na^+ uptake by plant cells. *Adv. Bot. Res*, *29*,

75-112.

Aragüés, R., et al. (1999). A new drip-irrigation system (DIS) for crop salt tolerance evaluation. *Soil Soc. Sci. Am. J*, *63*, 1397-1404.

Aragüés, R., et al. (1992). Evaluation of a triple line source sprinkler system for salinity crop production studies. *Soil Sci. Soc. Am. J*, *56*, 377-383.

Aragüés, R., et al. (1994). Foliar uptake of sodium and chloride in barley sprinkler-irrigated with saline water: effect of pre-irrigation with fresh water. *Eur. J. Agron*, *3*, 9-16.

Aslam, M., et al. (1984). Early effects of salinity on nitrate assimilation in barley seedlings. *Plant Physiol*, *76*, 321-325.

Ball, M. C. (1988). Salinity tolerance in the mangroves *Aegiceras corniculatum* and *Avicennia marina*. I. Water use in relation to growth, carbon partitioning and salt balance. *Aust. J. Plant Physiol*, *15*, 447-464.

Ball, M. C., et al. (1987). Salinity-induced potassium deficiency causes loss of functional photosystem II in leaves of the grey mangrove, *Avicennia marina*, through depletion of the atrazine-binding polypeptide. *Aust. J. Plant Physiol*, *14*, 351-361.

Barrett-Lennard, E. G. (1986). Effects of waterlogging on growth and NaCl uptake by vascular plants under saline conditions. *Recl. Reveg. Res*, *5*, 245-261.

Barrett-Lennard, E. G. (2003). The interaction between waterlogging and salinity in higher plants: causes, consequences and implications. *Plant Soil*, *253*, 35-54.

Barrett-Lennard, E. G., et al. (1999). The developing pattern of damage in wheat (*Triticum aestivum* L.) due to the combined stresses of salinity and hypoxia: experiments under controlled conditions suggest a methodology for plant selection. *Aust. J. Agric. Res*, *50*, 129-136.

Beatty, K. D., & Ehlig, C. F. (1973). A technique for testing and selecting for salt tolerance in sugar beet. *J. Am. Soc. Sugar Beet Tech*, *17*, 295-299.

Benes, S. E., et al. (1996). Foliar and root absorption of Na and Cl in maize and barley: implications for salt tolerance and use of saline sprinkler irrigation. *Plant Soil*, *180*, 75-86.

Bernstein, L. (1965). *Salt Tolerance of Fruit Crops*, USDA Inf. Bull. 292, US Govt. Printing Office, Washington DC, USA, 8.

Bernstein, L., et al. (1969). Effect of grape rootstocks on chloride accumulation in leaves. *J. Am. Soc. Hort. Sci*, *94*, 584-590.

Berthomieu, P., et al. (2003). Functional analysis of *AtHKT1* in *Arabidopsis* shows that Na^+ recirculation by the phloem is crucial for salt tolerance. *EMBO J*, *22*, 2004-2014.

Bower, C. A. (1959). *The Chemical Amendments for Improving Sodium Soils*, USDA Info. Bull. 195, US Govt. Printing Office, Washington DC, USA.

Byrt, C. S., et al. (2007). KLT1;5-like transporters linked to Na^+ exclusion loci in wheat, *Nax2* and *Kna1*. *Plant Physiol*, *143*, 1918-1928.

Causse, M. A., et al. (1994). Saturated molecular map of rice genome based on an interspecific backcross population. *Genet*, *138*, 1251-1274.

Chen, Z., et al. (2005). Screening plants for salt tolerance by measuring K^+ flux: a case study for barley. *Plant Cell Environ*, *28*, 230-246.

Colmer, T. D., et al. (1996). Interactive effects of Ca^{2+} and NaCl salinity on the ionic relations and proline accumulation in the primary root tip of *Sorghum bicolor*. *Physiol. Plant*, *97*, 421-424.

Colmer, T. D., et al. (2005). Improving salt tolerance of wheat and barley: future prospects. *Aust. J. Exp. Agric*, *45*, 1425-1443.

Cramer, G. R., et al. (1994). Salt tolerance is not associated with the sodium accumulation of two maize hybrids. *Aust. J. Plant. Physiol*, *21*, 675-692.

Cramer, G. R., et al. (1988). Kinetics of root elongation of maize in response to short-term exposure to NaCl and elevated Ca concentration. *J. Exp. Bot*, *39*, 1513-1522.

CSSRI, Vision 2020. (1997). *CSSRI Perspective Plan*, Central Soil Salinity Research Institute, Karnal, Haryana, India, 95.

Davenport, R., et al. (2005). Control of sodium transport in durum wheat. *Plant Physiol*, *137*, 807-818.

Downton, W. J. S. (1978). Growth and flowering in salt-stressed avocado trees. *Aust. J. Agric. Res*, *29*, 523-534.

Dubcovsky, J., et al. (1996). Mapping of the K^+/Na^+ discrimination locus *Kna1* in wheat. *Theor. Appl. Gen*, *92*, 448-454.

Dvořák, J., & Gorham, J. (1992). Methodology of gene transfer by homoeologous recombination into *Triticum turgidum*: transfer of K^+/Na^+ discrimination from *T. aestivum*. *Genome*, *30*, 639-646.

Dvořák, J., et al. (1988). On the evolution and the adaptation of Lophopyrum elongatum to growth in saline environments. *Proc. Nat. Acad. Sci*, *85*, 3805-3809.

El Hendawy, S. E., et al. (2005). Growth, ion content, gas exchange, and water relations of wheat genotypes differing in salt tolerances. *Aust. J. Agric. Res*, *56*, 123-134.

Flowers, T. J. (2004). Improving crop salt tolerance. *J. Exp. Bot*, *55*, 307-319.

Flowers, T. J., et al. (1986). Halophytes. *Q. Rev. Biol*, *61*, 313-337.

Flowers, T. J., et al. (2000). QTL: their place in engineering tolerance of rice to salinity. *J. Exp. Bot*, *51, 99-106.*

Flowers, T. J., et al. (1977). The mechanism of salt tolerance in halophytes. *Ann. Rev. Plant Physiol*, *28*, 183121.

Foolad, M. R., & Lin, G. Y. (1997). Absence of a genetic relationship between salt tolerance during seed germination and vegetative growth in tomato. *Plant Breed*, *116*, 363-367.

Forster, B. P., et al. (1988). Salt tolerance of two wheat–*Agropyron junceum* disomic addition lines. *Genome*, *30*, 559-564.

Fortmeier, R., & Schubert, S. (1995). Salt tolerance of maize (*Zea mays* L.): the role of sodium exclusion. *Plant Cell Environ*, *18*, 1041-1047.

Francois, L. E., & Maas, E. V. (1994). Crop response and management on salt-affected soils, In M. Pessarakli (Eds.), *Handbook of Plant and Crop Stress*, *Marcel Dekker*, New York, 149-181.

Frensch, J., & Hsiao, T. C. (1994). Transient responses of cell turgor and growth of maize roots as affected by changes in water potential. *Plant Physiol*, *104*, 247-254.

Frensch, J., & Hsiao, T. C. (1995). Rapid response of the yield threshold and turgor regulation during adjustment of root growth to water stress in *Zea mays*. *Plant Physiol*, *108*, 303-312.

Garcia, A., et al. (1997). The effects of selection for sodium transport and of selection for agronomic characteristics upon salt resistance in rice (*Oryza sativa* L.). *Theor. Appl. Gen*, *90*, 1106-1111.

Genc, Y., et al. (2007). Reassessment of tissue Na^+ concentration as a criterion for salinity tolerance in bread wheat. *Plant Cell Environ*, *30*, 1486-1498.

Ghassemi, F., et al. (1995). *Salinization of Land and Water Resources: Human Causes*, Extent, Management and Case Studies, CAB International, Wallingford, UK.

Gorham, J. (1994). Salt tolerance in the *Triticeae*: K/Na discrimination in some perennial wheatgrasses and their amphiploids with wheat. *J. Exp. Bot*, *45*, 441-447.

Gorham, J., et al. (1987). Chromosomal location of a K/Na discrimination character in the D genome of wheat. *Theor. Appl. Gen*, *74*, 584-588.

Gorham, J., et al. (1994). Varietal differences in sodium uptake in barley cultivars exposed to soil differences or salt spray. *J. Exp. Bot*, *45*, 895-901.

Gorham, J., et al. (1990). Partial characterisation of the trait for enhanced K^+-Na^+ discrimination in the D genome of wheat. *Planta*, *180*, 590-597.

Greenway, H., & Munns, R. (1980). Mechanisms of salt tolerance in non-halophytes. *Ann. Rev. Plant Physiol*, *31*, 149-190.

Greenway, H., & Osmond, C. B. (1972). Salt responses of enzymes from species differing in salt tolerance. *Plant Physiol*, *49*, 256-259.

Gregorio, G. B., et al. (2002). Progress in breeding for salinity tolerance and associated abiotic stresses in rice. *Field Crops Res*, *76*, 91-101.

Grieve, C. M., et al. (1994). Salinity affects the timing of phasic development in spring wheat. *Crop Sci*, *34*, 1544-1549.

Haro, R., et al. (2005). *HKT1* mediates sodium uniport in roots. Pitfalls in the expression of *HKT1* in yeast. *Plant Physiol*, *139*, 1495-1506.

Hasegawa, P. M., et al. (2000). Plant cellular and molecular responses to high salinity. *Ann. Rev. Plant Physiol. Mol. Biol*, *51*, 463-499.

Hollington, P. A., et al. (1999). *Assessment and development of salinity, sodicity and waterlogging tolerant wheat genotypes for India and Pakistan*, First Annual Report EU INCO-DC Project ERBIC 18CT 980305, Centre for Arid Zone Studies, University of Wales, Bangor, 126.

Hsiao, T. C., & Xu, L. K. (2000). Sensitivity of growth of roots versus leaves to water stress: biophysical analysis and relation to water transport. *J. Exp. Bot*, *51*, 1595-1616.

Hu, Y., & Schmidhalter, U. (1998). Spatial distributions and net deposition rates of mineral elements in the elongating wheat (*Triticum aestivum* L.) leaf under saline soil conditions. *Planta*, *204*, 212-219.

Huang, S., et al. (2008). Comparative mapping of *HKT* genes in wheat, barley, and rice, key determinants of Na^+ transport, and salt tolerance. *J. Exp. Bot*, *59*, 927-937.

Huang, S., et al. (2006). A sodium transporter (*HKT7*) is a candidate for *Nax1*, a gene for salt tolerance in durum wheat. *Plant Physiol*, *142*, 1718-1727.

IRRI. (1998). *Programme Report for 1997*, International Rice Research Institute, Los Baños, Philippines.

Isla, R., et al. (2003). Spatial variability of salt-affected soils in the middle Ebro valley (Spain) and implications in plant breeding for increased productivity. *Euphytica*, *134*, 325-334.

Isla, R., et al. (1997). Field screening of barley cultivars to soil salinity using a sprinkler and a drip irrigation system. *Plant Soil*, *197*, 105-117.

James, R. A., et al. (2006). Physiological characterization of two genes for Na^+ exclusion in wheat: *Nax1* and *Nax2*. *Plant Physiol*, *142*, 1537-1547.

James, R. A., et al. (2002). Factors affecting CO_2 assimilation, leaf injury and growth in salt-stressed durum wheat. *Func. Plant Biol*, *29*, 1393-1403.

James, R. A., et al. (2008). Genetic variation in tolerance to the osmotic stress component of salinity stress in durum wheat. *Funct. Plant Biol*, *35*, 111-123.

Jeschke, W. D. (1984). K^+-Na^+ exchange at cellular membranes, intracellular compartmentation of cations, and salt tolerance, In Staples R. C., & Toenniessen, G. H. (Eds.), *Salinity Tolerance in Plants: Strategies for Crop Improvement*, John Wiley and Co., New York, USA, 37-66.

Jeschke, W. D., et al. (1986). Effects of NaCl on ion relations and carbohydrate status of roots and on osmotic regulation of roots and shoots of *Atriplex Amnicola*. *Plant Cell Environ*, *9*, 559-569.

Jiang, Q., et al. (2006). Stomatal conductance is a key parameter to assess limitations to photosynthesis and growth potential in barley genotypes. *Plant Biol*, *8*, 515-521.

John, C. D., et al., (1977). Interaction of salinity and anaerobiosis in barley and rice. *J. Exp. Bot*, *28*, 133-141.

Jones, R. G. W., & Storey, R. (1978). Salt stress and comparative physiology in the Gramineae. II. Glycinebetaine and proline accumulation in two salt- and water-stressed barley cultivars. *Aust. J. Plant Physiol*, *5*, 817-829.

Kent, L. M., & Lauchli, A. (1985). Germination and seedling growth of cotton: salinity-calcium interactions. *Plant Cell Environ*, *8*, 155-159.

Khatun, S., & Flowers, T. J. (1995). Effects of salinity on seed set in rice. *Plant Cell Environ*, *18*, 61-67.

Koebner, R. M. D., et al. (1996). Responses to salt stress controlled by the homoeologous group 5 chromosomes of hexaploid wheat. *Plant Breed*, *115*, 81-84.

Koyama, M. L., et al. (2001). Quantitative trait loci for component physiological traits determining salt tolerance in rice. *Plant Physiol*, *125*, 406-422.

Kriedemann, P. E., & Sands, R. (1984). Salt resistance and adaptation to root-zone hypoxia in sunflower. *Aust. J. Plant Physiol*, *11*, 287-301.

Kurth, E., et al. (1986). Resistance of fully imbibed tomato seeds to very high salinities. *Plant Cell Environ*, *9*, 667-676.

LaHaye, P. A., & Epstein, E. (1969). Salt tolerance by plants: enhancement with calcium. *Science*, *166*, 395-396.

Laüchli, A. (1984). Salt exclusion: an adaptation of legumes for crops and pastures under saline conditions, In Staples, R. C., & Toenniessen, G. H. (Eds.), *Salinity Tolerance in Plants: Strategies for Crop Improvement*, John Wiley and Co., New York, USA, 171-187.

Laüchli, A., & Epstein, E. (1970). Transport of potassium and rubidium in plant roots, the significance of calcium. *Plant Physiol*, *45*, 639-641.

Laüchli, A., & Epstein, E. (1990). Plant responses to saline and sodic conditions (Chapter 6), In Tanji, K. K. (Eds.), *Agricultural Salinity Assessment and Management*, ACSE Manuals and reports on engineering practice No. 71, ASCE, New York, USA, 113-137.

Laüchli, A., & Wieneke, J. (1979). Studies on growth and distribution of Na^+, K^+ and Cl^- in soybean varieties differing in salt tolerance, Z. Pflanzennähr. *Bodenkunde*, *142*, 3-13.

Lin, H. X., et al. (2004). QTLs for Na^+ and K^+ uptake of the shoots and roots controlling rice salt tolerance. *Theor. Appl. Genet*, *108*, 253-260.

Lindsay, M. P., et al. (2004). A locus for sodium exclusion (*Nax1*), a trait for salt tolerance, mapped in durum wheat. *Funct. Plant Biol*, *31*, 1105-1114.

Lynch, J., et al. (1982). Na^+-K^+ relationships in salt-stressed barley, In Scaife A. (Eds.), *Plant*

Nutrition 1982, Proc. 9th Int. Plant Nutrition Colloq., CAB, Slough, UK, 347-352.

Maas, E. V. (1984). Salt tolerance of plants, In Christie, B. R. (Eds.), *Handbook of Plant Science in Agriculture*, CRC Press, Boca Raton, Florida, USA.

Maas, E. V. (1990). Crop salt tolerance (Chapter 13), In Tanji, K. K. (Eds.), *Agricultural Salinity Assessment and Management*, ACSE Manuals and reports on engineering practice No. 71, ASCE, New York, USA, 262-304.

Maas, E. V., & Grattan, S. R. (1999). Crop yields as affected by salinity, In Skaggs, R. W., & van Schilfgaarde, K. (Eds.), *Agricultural Drainage*, US Salinity Lab., Riverside, California, USA, 55-108.

Maas, E. V., & Grieve, C. M. (1990). Spike and leaf development in salt-stressed wheat. *Crop Sci*, *30*, 1309-1313.

Maas, E. V., & Hoffman, G. J. (1977). Crop salt tolerance - current assessment. *J. Irrig. Drain. Div., Am. Soc. Civ. Eng*, *103*(IR2), 115-134.

Maas, E. V., et al. (1983). Salt sensitivity of corn at various growth stages. *Irrig. Sci*, *4*, 45-57.

Mahmood, A., & Quarrie, S. A. (1993). Effects of salinity on growth, ionic relations and physiological traits of wheat, disomic addition lines from *Thinopyrum bessarabicum*, and two amphiploids. *Plant Breed*, *110*, 265-276.

Marcar, N. E. (1993). Waterlogging modifies growth, water use, and ion concentrations in seedlings of salt-treated *Eucalyptus camaldulensis*, *E. tereticornis*, *E. robusta* and *E. globulus*. *Aust. J. Plant Physiol*, *20*, 1-13.

Martin, P. K., et al. (1993). The effect of photoperiod insensitivity on the salt tolerance of amphiploids between bread wheat (*Triticum aestivum*) and sand couch grass (*Thinopyrum bessarabicum*). *Plant Breed*, *111*, 283-289.

Munns, R. (1985). Na^+, K^+ and Cl^- in xylem sap flowing to shoots of NaCl-treated barley. *J. Exp. Bot*, *36*, 1032-1042.

Munns, R. (1993). Physiological processes limiting plant growth in saline soil: some dogmas and hypotheses. *Plant Cell Environ*, *16*, 15-24.

Munns, R. (2002). Comparative physiology of salt and water stress. *Plant Cell Environ*, *25*, 239-250.

Munns, R., & Sharp, R. E. (1993). Involvement of abscisic acid in controlling plant growth in soils of low water potential. *Aust. J. Plant Physiol*, *20*, 425-437.

Munns, R., & Tester, M. (2008). Mechanisms of salinity tolerance. *Ann. Rev. Plant. Biol*, *59*, 651-681.

Munns, R., et al. (1983). Halo-tolerant Eukaryotes, In Lange, O., et al. (Eds.), *Physiological Plant Ecology: II. Responses to the Chemical and Biological Environment*, Encyclopaedia of Plant

Physiology, New Series Vol 12C, Springer Verlag, Berlin, Germany, 59-135.

Munns, R., et al. (2000). Genetic variation for improving the salt tolerance of durum wheat. *Aust. J. Agric. Res*, *51*, 69-74.

Munns, R., et al. (2003). Genetic control of sodium exclusion in durum wheat. *Aust. J. Agric. Res*, *54*, 627-635.

Munns, R., et al. (1995). The significance of a two-phase growth response to salinity in wheat and barley. *Aust. J. Plant Physiol*, *22*, 561-569.

Norlyn, J., & Epstein, E. (1984). Variability in salt tolerance of four triticale lines at germination and emergence. *Crop Sci*, *24*, 1090-1092.

Omielan, J. A., et al. (1991). Salt tolerance and ionic relations of wheat as affected by individual chromosomes of salt-tolerant *Lophopyrum elongatum*. *Genome*, *34*, 961-974.

Passioura, J. B., & Munns, R. (2000). Rapid environmental changes that affect leaf water status induce transient surges or pauses in leaf expansion rate. *Aust. J. Plant Physiol*, *27*, 941-948.

Pasternak, D., & De Malach, Y. (1994). Crop irrigation with saline water, In Pessarakli, M. (Eds.), *Handbook of Plant and Crop Stress*, Marcel Decker, New York, USA, 599-622.

Pitman, M. G. (1984). Transport across the root and shoot/root interactions, In Staples, R. C., & Toenniessen, G. H. (Eds.), *Salinity Tolerance in Plants: Strategies for Crop Improvement*, John Wiley and Co., New York, USA, 92-123.

Quarrie, S. A., & Mahmood, A. (1993). *Improving salt tolerance in hexaploid wheat, Annual Report 1992*, AFRC Institute of Plant Science, Research Cambridge Laboratory, John Innes Institute Nitrogen Fixation Laboratory and Sainsbury Laboratory, UK, 4.

Quarrie, S. A., et al. (2005). A high-density genetic map of hexaploid wheat (*Triticum aestivum* L.) from the cross Chinese Spring × SQ1 and its use to compare QTLs for grain yield across a range of environments. *Theor. Appl. Genet*, *110*, 865-880.

Qureshi, R. H., & Barrett-Lennard, E. G. (1998). *Saline Agriculture for Irrigated Land in Pakistan: a Handbook, ACIAR Monograph 50*, ACIAR, Canberra, Australia, 142.

Rains, D. W., et al. (1987). Saline drainage water reuse in a cotton rotation system, *Cal. Agric*, *41*(9-10), 24-26.

Raven, J. A. (1985). Regulation of pH and generation of osmolarity in vascular plants: a cost-benefit analysis in relation to efficiency of use of energy, nitrogen and water. *New Phytol*, *101*, 25-77.

Ren, Z. H., et al. (2005). A rice quantitative trait locus for salt tolerance encodes a sodium transporter. *Nat. Genet*, *37*, 1029-30.

Reynolds, M. P., et al. (2005). Prospects for utilising plant-adaptive mechanisms to improve wheat and other crops in drought- and salinity-prone environments. *Ann. Appl. Biol*, *146*,

239-259.

Royo, A., & Aragüés, R. (1993). Validation of salinity crop production functions obtained with the triple line source sprinkler system. *Agron. J*, *85*, 795-800.

Royo, A., & Aragüés, R. (1999). Salinity-yield response functions of barley genotypes assessed with a triple line source sprinkler. *Plant Soil*, *209*, 9-20.

Royo, A., et al. (1991). Descripcion y evaluacion de cuatro modelos de respuesta de cultivares de cebada a la salinidad. *Invest. Agr. Prod. Veg*, *6*, 319-330.

Royo, A., et al. (2000). Salinity-grain yield response functions of barley cultivars assessed with a drip-injection sprinkler system. *Soil Sci. Soc. Am. J*, *64*, 359-365.

Samad, A., et al. (2001). Waterlogging tolerance, In Reynolds, M. P., et al. (Eds.), *Application of Physiology in Wheat Breeding*, CIMMYT, Mexico City, Mexico.

Schubert, S., & Laüchli, A. (1986). Na^+ exclusion, H^+ release, and growth of two different maize cultivars under NaCl salinity. *J. Plant Physiol*, *126*, 145-154.

Semikhodskii, A. G., et al. (1997). Mapping quantitative trait loci for salinity responses in wheat, In Jevtic, S., & Pekic, S. (Eds.), *Drought and Plant Production Vol. 2*, Proc. Int. Symp., Donji Milanovac, Serbia, Sept 1996, Agricultural Research Institute, Belgrade, Serbia, 83-92.

Shannon, M. C. (1997). Adaptation of plants to salinity. *Adv. Agron*, *60*, 75-120.

Shannon, M. C., et al. (1994). Whole-plant response to salinity, In Wilkinson, R.E, (Eds.), *Plant-Environment Interactions*, M. Dekker, New York, 199-244.

Storey, R., & Walker, R. R. (1999). Citrus and salinity, *Sci. Hort*, *78*, 39-81.

Taeb, M., et al. (1992). Association between genes controlling flowering time and shoot sodium accumulation in the *Triticeae*. *Plant Soil*, *146*, 117-121.

Takehisa, H., et al. (2004). Identification of quantitative trait loci for plant growth of rice in paddy field flooded with salt water. *Field Crops Res*, *89*, 85-95.

Tanji, K. K., & Kielen, N. (2002). *Agricultural Drainage Water Management in Arid and Semi-arid Areas*, FAO Irrig. Drainage Paper 61, FAO, Rome, Italy, 188.

Termaat, A., et al. (1985). Shoot turgor does not limit shoot growth of NaCl-affected wheat and barley. *Plant Physiol*, *77*, 869-872.

van Genuchten, M. T. (1983). *Analyzing Crop Salt Tolerance Data: Model Description and Users' Manual*, USDA-ARS US Salinity Lab. Res. Rep. No. 120, Salt model computer program available at http://ars.usda.gov/Services/docs.htm?docid=8957.

van Genuchten, M. T., & Hoffman, G. J. (1984). Analysis of crop salt tolerance data, In Shainberg, I., & Shalhevet, J. (Eds.), *Soil Salinity under Irrigation - Process and Management*, Ecological Studies 51, Springer Verlag, Berlin, Germany, 258-271.

Yeo, A. R., et al. (1990). Screening of rice (*Oryza sativa* L.) genotypes for physiological characters contributing to salinity resistance, and their relationship to overall performance. *Theor. Appl. Genet*, *79*, 377-384.

Yeo, A. R., et al. (1991). Short and long-term effects of salinity on leaf growth in rice (*Oryza sativa* L.). *J. Exp. Bot*, *42*, 881-889.

Zhong, G. Y., & Dvořák, J. (1995a). Evidence for common genetic mechanisms controlling the tolerance of sudden salt stress in the tribe *Triticeae*. *Plant Breed*, *114*, 297-302.

Zhong, G. Y., & Dvořák, J. (1995b). Chromosomal control of the tolerance of gradually and suddenly imposed salt stress in the *Lophopyrum elongatum* and wheat, *Triticum aestivum* L., genomes. *Theor. Appl. Genet*, *90*, 229-236.

CHAPTER 4.

ADVANCES IN BREEDING WHEAT FOR SALT TOLERANCE

Luxiang Liu*, Huijun Guo, Linshu Zhao, Jiayu Gu and Shirong Zhao

National Key Facility for Crop Gene Resources and Genetic Improvement, Institute of Crop Sciences, Chinese Academy of Agricultural Sciences, Beijing 100081, China

*Correspondence: luxiang@263.net.cn

1. INTRODUCTION

Salinization and drought are main abiotic stresses that seriously affect the growth and development of cultivated plants and agricultural production worldwide. Over 7% of the world's land is affected by either salinity or sodicity. Food production and the welfare of the people living in the arid and saline areas are severely affected. In China the estimated total area of saline-alkali soils is about 33.5 Mha, from which 6.7 Mha have been reclaimed and brought under cultivation and approximately 20 Mha of saline-alkali land are waiting to be reclaimed. In addition more than 6.7 Mha of fertile farmland (about 10% of the total land area under cultivation in China) are affected by secondary salinization due to the inappropriate irrigation methods.

Wheat is one of the most important world food crops, and its productivity directly affects human survival and quality of life. In China wheat is now the second most important crop with more than 26 Mha of land cultivated annually. There is a growing concern for the genetic improvement of the salt tolerance and drought stress resistance in wheat and these topics are receiving increasing attention by the breeders. Salt and drought tolerance are quantitative characters. There is a shortage in available germplasm that can be utilized. Salt-tolerant variety development by conventional methods has been generally limited. The development and achievements of the nuclear technology and related biotechnology provide the opportunity to tackle this difficult breeding problem effectively. This chapter will review the current state and future prospects for the improvement of salt-tolerant wheat varieties, in particular through the use of mutation and biotechnology techniques.

2. SOME BASIC PHYSIOLOGICAL PARAMETERS OF SALT TOLERANCE

Plant salt tolerance is generally defined as the inherent ability of the plant to withstand the effects of high concentration of salts in the root or the plant's leaves without significant adverse effects on plant productivity. Salt tolerance is a complex trait involving responses to cellular ionic and osmotic stresses and their consequent metabolic effects such as oxidative stress and whole plant coordination. These developments have been summarized in excellent reviews (Flowers et al., 1977; Greenway and Munns, 1980; Hasegawa et al., 2000; Munns and Tester, 2008; Pang and Wang, 2008).

Despite intensive scientific efforts made over the last decade, the physiological processes and mechanisms involved in plant sensitivity and tolerance to salinity are still not well understood due to their complexity. Owing to the lack of better options, the identification of traits and genes for genetic manipulation towards increased salt tolerance relies on unproved dogmas about tolerance mechanisms.

By far, the problems in designing proper selection criteria for salinity tolerance are a major drawback in making progress in breeding work. Nevertheless some criteria have been proposed and employed for salt tolerance screening in breeding programs or for mechanism studies. Zhao and Dou (1998) reviewed more than ten criteria used in salt tolerance identification and screening in wheat. Useful plant morphological and growth indicators under salt stress are seedling survival ratio, root length, shoot height and seedling weight, tiller number in spring and grain yield. Among the physiological and biochemical criteria that can be used under salt stress conditions are K^+/Na^+ ratio in leaves, cell membrane permeability, ATP and superoxide dismutase (SOD) enzyme activities, glycinebetaine and polyamine accumulation, cell wall hydroxyproline and cell surface glycoprotein content, and salinity stress-induced tissue-specific protein. Munns et al. (2002) recommended rates of Na^+ or Cl^- accumulation in leaves, degree of leaf injury, seedling root length, and germination percentage as the key traits for salt tolerance screening in wheat.

Seed germination in saline media is often used as a single criterion or in combination with other criteria. This criterion was found practical and effective to identify material tolerant to salinity at this stage (Blum, 1988). However, plants may be more tolerant or susceptible to salinity at germination than at subsequent growth stages (Noble, 1983). It has, therefore, been proposed that the use of germination in saline media is an inappropriate criterion for species more tolerant to salinity at germination because the problem lies in subsequent susceptible stages. On the other hand, Noble (1983) suggested that selection at germination is important in species that are relatively susceptible at this stage.

Selection during vegetative growth may involve several criteria. Seedling or plant dry matter as an expression of total growth is an important criterion that integrates the various possible effects of response to salinity into one measure of resistance. Wheat plants grown at 100 mmol L^{-1} NaCl produced less dry matter and chlorophyll content than those without NaCl, and supplementary Si at both 0.25 mmol L^{-1} and 0.5 mmol L^{-1} ameliorated the negative effects of salinity on plant dry matter and chlorophyll content (Tuna et al., 2008). There are, however, several problems related to this growth measurement as follows: (i) it is a destructive method, which is unacceptable for selection at early generations; and (ii) genotypes may differ in their growth potential, and their growth differences under salinity stress may represent their potential growth capacity rather than their specific tolerance.

As for the case of several other environmental stresses, the growth of the genotypes in the study under stress can be used as a criterion of tolerance only if it is compared with growth under non-stress conditions. However, the measurement of the rate of growth reduction under two stress environments entails double work load, costs and facilities.

Other criteria could be used as indirect estimates of the effect of salinity during vegetative growth. Leaf desiccation and death caused by the salinity stress have been assessed either in terms of leaf numbers affected or by a visual estimate of the affected leaf area in proportion to total leaf area (Noble and Rogers, 1992). In extreme cases the situation is simple enough, where the population clearly truncates into surviving and dead plants (Blum, 1988). In species where salinity tolerance is conditioned largely by salt exclusion, leaf analysis for the specific ion content would seem to be an important criterion (Yeo and Flowers, 1983).

Salt tolerance in the *Triticeae* is associated with Na^+ exclusion, which limits the entry of sodium into the plant and its transport to leaves. Na^+ exclusion from the transpiration stream reaching the leaves is controlled at three stages: (i) selectivity of the root cells taking up cations from the soil solution; (ii) selectivity in the loading of cations into the xylem vessels in the roots; and (iii) removal of sodium from the xylem in the upper part of the roots and the lower part of the shoot (Munns et al., 2002).

In wheat, salt tolerance is associated with low rates of transport of Na^+ to shoots with high selectivity for K^+ over Na^+ (Ding et al., 2006). The most successful criterion relates to rates of Na^+ or Cl^- accumulation in leaves, measured as the increase in salt in a given leaf over a fixed period of time. Bread wheat (hexaploid) cultivars are able to exclude Na^+ from the leaves but durum wheat (tetraploid) cultivars lack this trait (Dubcovsky et al., 1996). Recently, a novel source of Na^+ exclusion was identified in a durum landrace (Munns et al., 2000). The landrace had very low rates of Na^+ accumulation in the leaf blade, as low as bread wheat cultivars, and maintained a high rate of K^+ accumulation, with consequent high K^+/Na^+ discrimination. The low Na^+ trait was shown to confer a significant yield advantage at moderate soil salinity (Husain et al., 2003), indicating that this novel germplasm provides the opportunity to improve the salt tolerance of cultivated durum wheat. The low rates of Na^+ uptake and accumulation in a given leaf as a non-destructive and accurate quantitative trait have been used in durum wheat (Munns et al., 2002). But the Na^+ and Cl^- exclusion did not always reflect the salt tolerance, whereas K^+ in the leaves and Ca^{2+} in the leaves and stems were closely associated with genotypic differences in salt tolerance (El-Hendawi et al., 2005). Significantly positive correlations among K^+/Na^+, reproductive growth period and total growth period were noted in salt-sensitive varieties, however, no significant relations appeared among those parameters in salt-tolerant varieties, indicating that higher salinity tolerance varieties of winter wheat could relieve senescence at the reproductive stage (Zheng et al., 2008).

Criteria related to the capacity for osmoregulation as a component for tolerance were suggested (Blum, 1988; Liang et al., 2006). Membrane stability has been widely used to differentiate stress-tolerant and sensitive cultivars (Blum and Ebercon, 1981) and usually higher membrane stability is correlated with abiotic stress tolerance (Premachandra et al., 1991). The cell membrane stability technique was proved to be suitable for screening wheat tolerance to high salinity and detecting differences that may arise due to cumulative effects of salinity and reduced water contents. The cell membrane stability as a salt-tolerant criterion could be used more effectively than grain yield for screening large quantities of germplasm at the seedling stage (Farooq and Azam, 2006).

As root growth often expresses well the relative tolerance of a plant to mineral toxicity, it was considered also as a possible criterion of tolerance to salinity (Blum, 1988). Due to the genetic basis and significant correlation with grain yield, root length could be used as a selection criterion to identify salt-tolerant wheat genotypes (Zulfiqar, 2004).

Physiological studies, geared toward identification of specific molecular/cellular processes of sensitivity or tolerance, might potentially shorten the duration required for the development of salt-tolerant crop species. Most crop species are affected by salt stress during their growth cycle. As a result, their growth and economic yield are substantially reduced under salt stress. They also show variation in sensitivity to salt stress during their stages of development. Salinity stress tolerance appears to be developmentally regulated, it is a stage-specific phenomenon; tolerance at one stage of plant development is often not correlated with tolerance at other developmental stages. Thus, specific ontogenic stages including seed germination and emergence, seedling survival and growth, and vegetative growth and reproduction, should be evaluated separately for assessment of tolerance and the identification, characterization and utilization of useful genetic components. Each developmental stage (which may be considered as a separate trait) may require a different screening procedure, and simultaneous or sequential screening may be impractical or impossible. Partitioning of salinity stress tolerance into component traits related to ontogenic stages may facilitate a better understanding of the genetic basis of stress tolerance and development of stress-tolerant plants (IAEA, 2002).

3. GENETIC DIVERSITY OF SALT TOLERANCE

The availability of a significant amount of variation in salinity tolerance in the gene pool is necessary for developing salt-tolerant crop genotypes. Norlyn (1980) and Shannon (1984) discussed the subject of genetic variability among plants in response to salinity. Genetic variation for salt tolerance has been found in world collections of bread wheat (Qureshi et al., 1980; Kingsbury and Epstein, 1984; Sayed, 1985). Singh and Chatrath (1992) screened and found variation for salt tolerance in tissue-culture-derived wheat lines developed at CIMMYT. The genes controlling tolerance are located on all seven homoeologous chromosome groups of the *Triticeae.* However, group 4 and group 5 chromosomes are predominant for most stresses (Forster, 1994).

Guo et al. (2001) screened 161 Tibetan wheat accessions for salt tolerance under three different salt concentrations and found that 9.5% of the accessions possessed high salt tolerance and three among them were salt-insensitive revealing the rich variation in salt tolerance in the Tibetan wheat germplasm. Zhu et al. (1996) studied 900 collections of wheat varieties or advanced lines and identified sixty-one accessions with better salt tolerance and agronomical characters. Ma and Weng (2005) screened 28 spring wheat varieties introduced from America for salt tolerance at both germination and seedling stages. Thirteen of them showed high salt tolerance with regard to grain yield and its components, and from them two varieties SW10 and SW12 showed higher performance and grain yield under high saline soil conditions.

Zulfiqar (2004) reported variability in the seedling responses of 98 wheat accessions to increasing NaCl concentrations in the rooting medium and indicated the potential for breeding salt-tolerant genotypes within this species. The genetic variation for salt tolerance was shown to be influenced predominantly by the genes with additive effects based on root length. The heritability estimated under low and high salinity was appreciable and seems to be promising for genetic improvement of salt tolerance in wheat.

Trethowan and Mujeeb-Kazi (2008) reviewed three sources of novel genetic variability, namely synthetic wheat, landrace cultivars, and alien introgressions as well as their applicability to wheat breeding for improving environmental stress performance. Synthetic hexaploid wheat derived by crossing tetraploid wheat with *Aegilops tauschii* provides new genetic variability for adaptation to abiotic stresses such as drought, high temperature, salinity, and waterlogging. Synthetic-derived materials have performed well in many stressed environments globally. There is significant unexploited variation among landraces and modern wheat cultivars to improve the stress adaptation of cultivated wheat. They concluded that there was sufficient genetic variation in the wheat gene pool to ensure the continued improvement of wheat adaptation to abiotic stresses.

4. CONVENTIONAL BREEDING FOR SALT TOLERANCE

The conventional breeding methods have been used to improve wheat salt tolerance in different countries such as India, Pakistan and China, etc. In India, efforts to improve salt tolerance of crops by scientists at CSSRI, Karnal have led to the development and release of varieties of wheat, rice, chickpea, and Indian mustard suitable for cultivation in alkali and saline soils (Sharma and Goyal, 2003). The first Indian salt-tolerant wheat variety KRL1–4, derived from a cross of the highly salt-tolerant Indian landrace Kharchia 65 with a popular high-yielding wheat variety WL 711, was released in 1990. Another salt-tolerant wheat variety, KRL 19, was released in 1999. It has better tolerance to rusts and other diseases and it is a safe replacement of KRL1–4. In India, almost all salt-tolerant wheat germplasm is derived from Kharchia 65 (Hollington, 1998). In Pakistan, LU26S and a series of salt-tolerant wheat varieties named after Saline Agriculture Research Cell (SARC-1, 2, 3 and 4) were also developed and released.

In China, a high salt-tolerant variety Jimai32, developed by the State Sino-Czechic Farm in Hebei province through a cross of a local high-yield selection Nongda311 and a Kefan68 derived from wide hybridization, was released in 1992. It is still currently cultivated in the saline areas of Hebei province for its high tolerance to salt and drought as well as good yield performance (Zhao et al., 2000). Using Jimai32 as the male parent to make a cross with a high-yield variety Lifen6145, the separating progenies were advanced along with the parallel alternative selection in two environments of salt stress and non-salt stress at the same time. This work led to the development of another variety Cang6001 with good tolerance to salt and drought stress, which was released in 1998 and planted as one of the best varieties up to now in the saline areas of Hebei province (Zhao et al., 2000). In the past decade, a recurrent selection via a *Taigu* genic male-sterile wheat line to pyramid different sources of salt tolerance genes has been employed in the wheat breeding programme for salt tolerance. Some promising advanced lines from the Cangzhou Academy of Agricultural Sciences such as Cang026, Cang030 and Cang036 have been put into multi-location regional trials for release in Hebei province (Liang et al., personal communication).

The Dezhou Academy of Agricultural Sciences in Shandong province is another institute specialized on crop salt tolerance improvement in China. Some elite salt-tolerant wheat varieties such as Dexuan1, Lumai10, Lude1 and Dehang961 have been successfully developed by using re-selection or cross-breeding methods and put into production since the 1960s.

There are also several salt-tolerant wheat varieties such as Changfangbai, Shidong5, Shidong7 and Xindong26, etc. that have been developed and officially released in Xinjiang since the 1980s. Both single cross and multi-cross between or among the salt-tolerant selections and local high-yielding varieties were usually made to upgrade the breeding efficiency for salt tolerance improvement (Luo and Ren, 2001; Luo et al., 2005; Luo et al., 2006).

5. WIDE HYBRIDIZATION IN WHEAT

Some extensive reviews on the use of wild relatives to improve the salt tolerance of wheat are available (Colmer et al., 2006; Munns and Richards, 2007). *Thinopyrum bessarabicum* is a perennial species within the graminaceous tribe *Triticeae* and is more salt-tolerant than the annual bread wheat (*Triticum aestivum* L.). The amphidiploid produced (Forster and Miller, 1985) by hybridizing the bread wheat variety Chinese Spring and *Thinopyrum bessarabicum* was found to be more tolerant in terms of survival and ability to produce grain at moderate salinity (250 mol m^{-3}) than the wheat varieties Chinese Spring or even Kharchia. The greater resistance of the amphidiploid was attributed to its inheritance of more efficient exclusion of Na^+ and Cl^- from younger leaves and reproductive tissue (Gorham et al., 1986).

It is also known that hexaploid bread wheat (AABBDD) expresses more K^+/Na^+ selectivity than tetraploid (AABB) wheats, as K^+/Na^+ was found to be associated with the chromosome 4 of the D genome (Gorham et al., 1987). Although there appears to be little allelic variation for this character in hexaploid wheat, variation may exist in *Aegilops squarrosa*, donor of the D genome, and other relatives of wheat carrying this genome. Variation in the salt tolerance of the D genome was shown to influence the salt tolerance of synthetic hexaploids. Schachtman et al. (1992) produced synthetic hexaploids from five *Aegilops squarrosa* accessions varying in salt tolerance and two salt-sensitive *T. turgidum*. The relative grain yield of the hexaploids in 150 mol m^{-3} NaC1 was greater than that of the tetraploid parents, primarily due to the maintenance of grain weight under salinity. *Aegilops tauschii* (DD) has been hybridized with durum wheat (AABB) to produce synthetic hexaploid wheat (Schachtman et al., 1992; Mujeeb-Kazi and Diaz de Leon, 2002).

Studies done in China showed that *Aegilops crassa* 6x cytoplasm was able to produce significant genetic effects on the salt tolerance of common wheat. The character and range of effect value were related to the nucleoplasm of the alloplasmic common wheat. The salt tolerance of the alloplasmic common wheat could be significantly increased in specific nucleoplasm combinations. The results from experiments of callus-inducing and development in salt culture-medium, seed germination in salt solution and seedling development in salt Hoagland solution indicated that the alloplasmic lines *Aegilops crassa* 6x-Jian 26 and *Aegilops crassa* 6x-SMH1694 were more salt-tolerant than their nuclear parents. The results at recovering stage and mature period demonstrated that some alloplasmic common wheats with improved nuclear genotypes were more salt-tolerant than or equal to the salt-tolerant control variety Keyi 26 (Hou et al., 2004).

Studies using wheat tetrasomic lines ($2x = 44$) and wheat/*Agropyron junceum* disomic lines ($2x = 44$) have shown that chromosomes 2A, 2B, and 2D of wheat and 2J of *Agropyron junceum* carry genes that confer salt susceptibility. However, chromosome 2J also appears to carry genes for salt tolerance (Forster et al., 1988). These findings suggest the existence of genes with major effects that might be exploited to increase salt tolerance in wheat.

A breeding programme was carried out in Pakistan to introduce salt tolerance from *Aegilops cylindrica* (CCDD) to hexaploid wheat (Farooq et al.,1992a; 1992b). The backcrossed lines produced from hybrids between *Aegilops cylindrica* and the Pakistani cultivars LU26 and Pak81 were tested in both saline and non-saline fields and were found to be both salt- and drought-tolerant (Farooq et al., 1995; Farooq, 2004). Wheat lines WL1076 and WL41 out-yielded LU26, the salt-tolerant parent, and required less irrigation water and fertilizer than other genotypes. They also had higher yields than the current wheat variety Inqlab (Farooq, 2004; Farooq and Azam, 2005).

Amphidiploid wheat plants were also produced by crossing two moderately salt-tolerant wheats, namely *Triticum turgidum durum* (AABB) and *Aegilops speltoides* subsp. *speltoides* (SS). They showed greater salinity tolerance than either parent assessed with the criterion of root growth in different NaCl concentrations (Noori, 2005).

Tall wheatgrasses (E or J genomes) are very salt-tolerant. At low to moderate salinity, they have a similar decline in biomass as it does barley and the more tolerant bread wheat varieties but they continue to grow at high salinity, even up to seawater concentrations and beyond (Munns and Richards, 2007).

King et al. (1996) assessed the salt tolerance of the wheat/*Thinopyrum bessarabicum* addition lines and 5Eb(5A) and 5Eb(5D) substitution lines. The 5Eb addition line survived better than the Chinese Spring variety while the substitution lines survived better than not only the Chinese Spring variety but also the addition line, showing that the gene or genes had a greater effect when substituted for an homoeologous wheat group 5 chromosome than when present as an additional chromosome. King et al. (1997) developed a new cereal (*Tritipyrum*) as it was triticale. The tetraploid wheats were hybridized with *Th. bessarabicum* and wheat-like amphiploids produced. They tested some of these materials with the octaploid amphiploid Chinese Spring × *Th. bessarabicum* and the wheat parents at four salinity levels. Survival was far better in the *Tritipyrum* lines at 150 mol m^{-3} NaCl, where 90% of the wheat died, but was much less at 200 mol m^{-3} and 250 mol m^{-3} although little or no wheat survived at these salinity levels.

Thinopyrum ponticum (decaploid, E genome) is a very salt-tolerant "tall wheatgrass" commonly used as a forage crop in saline land. Somatic hybridization techniques were used to transfer *Thinopyrum ponticum* chromosomes into bread wheat and the introgression lines derived from somatic hybrids between *Thinopyrum ponticum* and *Triticum aestivum* L. cv. Jinan 177 were screened for salt tolerance in hydroponic experiments. Their growth rates, salt tolerance index and the content of free proline, Na^+, K^+ and Na^+/K^+ ratios in the leaf were compared with those of wheat cv. Jinan 177. The F_4 and F_5 generation lines were tested under natural saline conditions in two locations. One line expressed higher salt tolerance than its parental wheat and a check salt-tolerant cultivar. Karyotype analysis showed that the two lines possessed 42 chromosomes of wheat introgressed with small chromosome segments from *Thinopyrum ponticum*. Thus, the salt tolerance of *Th. ponticum* appears to have been introgressed into bread wheat with the *Thinopyrum ponticum* chromatin stably inherited (Chen et al., 2004). Salt-tolerant bread wheat lines developed from a translocated segment of *Thinopyrum junceum* has been recently registered in the USA as W4909 and W4910 (Wang et al., 2003).

6. INDUCED MUTATIONS FOR SALT TOLERANCE IMPROVEMENT IN WHEAT

Induced mutations have been used mainly to improve particular characters in well-adapted local varieties or to generate variation difficult to be found in germplasm collections. The high efficiency of classical mutagenesis to generate mutations valuable for breeders has been widely proven and documented through the official release of 2,570 mutant varieties including, as indicated in the FAO/IAEA Mutant Varieties Database (http://wwwinfocris.iaea.org/MVD/). New varieties with increased resistance to disease, salinity, high and low temperatures and lodging have been developed by induced mutation with physical and chemical mutagens (Jain, 2001). From a general survey of the research progress made over the past years, the mutation induction approaches employed for improvement of salt tolerance of wheat and other crops can be summarized as follows: (i) induction and selection by radiation of seed; (ii) *in vitro* selection through tissue culture-derived variation; and (iii) indirect utilization or cross of mutants for salt-tolerant improvement (Liu et al., 2003a).

6.1 Mutation induction from seed treatment

Seeds are the most commonly used materials for induced mutation. Mutagenic treatments fall into two main classes: physical (γ rays, X rays, laser, UV, etc.) and chemical (EMS and NaN_3, etc.). Once the seed of a parental line has been treated with a mutagen it is referred to as M_0 seed and produces M_1 plants. Because of the possible chimeras and physiological disorders, normally selection is not practiced until later generations. Generally, individual mutations could be screened and selected in the M_2 or M_3 population. For the targeted trait of salt tolerance, mutation screening and selection of mutants should be conducted under the certain salt stress pressure. This method has been employed in Chinese wheat improvement programs since the 1970's. Some wheat mutant varieties with improved salt tolerance have been developed, officially released, and put into production. For instance, the mutant variety Jiaxuan No.1 was examined and approved in 1974, Changwei No.19 in 1978 and Yuandong No. 3 in 1986 (Liu et al., 2003a).

Guo et al. (1997) obtained saline-tolerant wheat lines by using 8.6 Gy X ray irradiation. In their experiment, germinating wheat seeds soaked in caffeine (1 mg mL^{-1}) for 3 h before irradiation or treated with heat (43℃, 20 min) after irradiation were cultured in NaCl solution with concentrations ranging from 1.4% to 2.0% for 15 days in order to select mutants for salinity tolerance. Wheat seeds were continuously grown in 1.9% and 2.0% NaCl solutions and selected for survival until M_5. Twelve stable saline-tolerant wheat selections were tested in the field. Two lines, i.e. Zhishen No.7 and Zhishen No.2, yielded twice as much than the control cv. Lumai No. 10 when planted in 0.6% saline soil.

Another successful example of induced mutation is the well-known barley variety Golden Promise, which is a γ rays-induced high-yielding and short-height mutant from the original variety Maythorpe. It was found that Golden Promise was more salt-tolerant than Maythorpe in terms of yield and shoot Na^+ because it was able to limit the amount of sodium taken up under salt stress. The main beneficial effect was derived from the reduction in the number of sterile spikelets in Golden Promise compared to Maythorpe (Forster et al., 1994; Foster, 2001). This mutant variety has been used as the parent of more than 150 leading barley varieties worldwide, and it has made a major impact on the brewery industry in Europe.

6.2 Tissue culture-derived salt tolerance

Among all environmental stresses, the possibilities of *in vitro* selection have been widely explored for salinity tolerance. This was due to the importance of the issue, the difficulties in achieving final results in plant selection, and the ease of applying salinity stress to any *in vitro* system (Blum, 1988). *In vitro* selection for salt tolerance commonly occurs as a result of a temporary adaptation; cells are able to compartment the excessive salts into vacuoles, and survive by adjusting the osmotic pressure. The first stage after the establishment of a cell culture from any suitable source material is the induction and isolation of salt-tolerant cells and cell lines in the culture. Salinity stress is commonly applied by the addition of NaCl to the culture. In order to eliminate variants that may have only some transient adaptation to the saline culture, salinity stress has to be high enough to kill more than 95% of the cells (Blum, 1988). But Barakat and Abdel-Latif (1996) also reported that the stepwise method of increasing NaCl in the medium was more effective for plant regeneration than other methods. The second stage involves regenerating plants from the salt-tolerant cell lines. Irrespective of salinity tolerance, plant regeneration from cell lines is a well-established routine in wheat. However, because of the genotype dependence, it is highly advised to establish varieties with high regeneration capacity before work in cell culture is initiated. A highly efficient regeneration system for *in vitro* culture of young spikes in wheat has been developed to decrease or overcome the genotype dependence (Liu et al., 2001).

As the regenerated plantlets may or may not carry salt tolerance, the salinity tolerance of the regenerated plants must be retained through plant reproduction. Dracup (1991) postulated that the selection of cultured cells may be more productive if focused on specific cell-based physiological traits (such as Na^+ accumulation, turgor regulation or tolerance to high Na^+/Ca^{2+} ratio) rather than tolerance to high NaCl concentration only. The successful production of salt-tolerant regenerated plants revealed that, in spite of the large differences in the conditions where the cultured cells are selected and the regenerated plants are evaluated, at least some of the cellular mechanisms of salt tolerance operating in cultured cells are also functional in the regenerated plants. In salt tolerance comparison studies of whole plants and isolated cells, a correlation between the mechanisms operating in cultured cells and the whole plant has been found (Tal, 1994).

Cell or tissue culture generates a wide range of genetic variation in plant species, which in turn can be incorporated into plant breeding programs. It could be further enhanced by applying various physical and chemical mutagen treatments. Combined with *in vitro* selection, mutants with useful agronomic traits, e.g. salt or drought tolerance or disease resistance, can be isolated in a relatively short period of time. This approach has been successfully employed in various crop species to obtain salt-tolerant cell lines (Jain, 2001; Hossain et al., 2006; Liu et al., 2003a; 2007).

The use of doubled haploid (DH) breeding through anther and microspore cultures saves time in producing many generations normally needed to obtain pure breeding lines. It also enhances the effectiveness of the selection of the desired recombinants. The advantages of DH technologies have long been recognized by the breeders. More than 300 varieties have been produced with the use of various DH methods in several crops (Szarejko and Forster, 2007). In China, pure breeding diploid lines have been developed from DH lines produced through anther and microspore *in vitro* culture. Nearly 100 new varieties officially released have been developed in crop plants such as wheat, rice, rapeseed and barley (Liu, unpublished data).

Mutagenesis in combination with DH-systems has been successful in developing valuable mutations for other characters such as salt- and drought-tolerant crop lines. Increasing certain physical/chemical mutagens and salt stress on the culture has become a very successful way to select salt-tolerant mutants of wheat.

A technological system for salt-tolerant wheat selection through the use of the DH technique combined with nuclear irradiation has been well established at the Institute of Crop Sciences, Chinese Academy of Agricultural Sciences (Liu et al., 2006). Salt-tolerant variants could be obtained via anther culture in salt-stressed media. Wheat anthers were irradiated with 1.5 Gy γ rays and then cultured on medium containing 0.5% NaCl for *callus* induction. After recurrent selection on NaCl-enriched medium, stable tolerant cultures were isolated and regenerated into plantlets for progeny test. A promising line H6756 with prominent salt- and drought-tolerant characters has been developed and officially released in 2004 in Shandong province. The average yield of H6756 in the regional multi-location test was 12.5% higher than that of the local control variety DK961. In dozens of demonstration trials conducted in Beijing, Shandong and Hebei provinces its yield potential was 6.75–7.50 t ha^{-1} in saline land areas with about 0.3%–0.4% NaCl content (Liu et al., 2007).

The process of selecting salt-tolerant varieties can be sped up effectively by using the salt-tolerant mutants as cross parents and combining with DH technology. The anthers of the F_1 generation of Yong445 ×Yuandong 3 were induced on a medium containing 0.4% NaCl for the first selection and then *callus* differentiation was conducted on the medium containing the same concentration of NaCl for the second selection. Regenerated pollen plants were grown for propagation in the soil without salt stress for one generation to remove possible existing physiological adaptability or post-natal genetic salt-tolerant phenotype. After three successive generations of field identification and selection for salt tolerance, the salt-tolerant line H89 was obtained. The field identification of the salt tolerance of H89 was confirmed during many years. The H89 line can be grown in saline soil containing up to 0.73% NaCl. In the salinized soil with concentrations ranging from 0.17% to 0.73% NaCl, the yield of H89 was superior to that of the local control variety Laizhou 953 (Zheng et al., 1996).

A series of salt-tolerant mutants have been also developed by using *calli* of anther or matured embryo after mutagenizing with ethyl methanesulfonate (EMS), Pingyangmycin (PYM) or Zhengdingmycin (ZDM). Salt-tolerant *calli* were introduced in an inducing medium containing a certain amount of mutagenic agent, and the salt-tolerant plants could be easily regenerated after such *calli* were transferred to the differential medium with the same amount of salt as the selective medium. Genetic analysis of the high generation materials of salt-tolerant wheat mutants showed that the salt-tolerant trait is not only controlled by nuclear genes but also affected by cytoplasmic factors (Shen et al., 1993; 1997; Zhao et al., 2007). Information on the stable salt-tolerant variants of wheat obtained from anther culture *in vitro* selection could be also found in the literature (Xiao et al., 1989; Zhao et al., 1994; 1995).

7. IDENTIFICATION OF CANDIDATE GENES FOR SALINITY TOLERANCE

A comprehensive discussion of the candidate genes for salt tolerance is given by Tester and Davenport (2003), with particular reference to roots and by Munns (2005) in relation to leaves and an expected phenotype. In salt-resistance studies, specific changes in the profile of proteins have been observed in many plants (Gulick and Dvorak, 1987; Caruso et al., 2008). As a result, new proteins could be synthesized or quantity changes induced under salt stress conditions. Several proteins have been identified to play a prominent role in response to salt stress. By using SDS-PAGE analysis, Zheng et al. (1994) and Zhu et al. (1996) observed significant increases of the proteins of 26 kDa and 52 kDa in the salt-tolerant wheat variety whereas such changes were much smaller in the salt-sensitive genotypes, implying that these two kinds of proteins may be related to wheat salt tolerance.

Huo et al. (2004) analyzed the proteome of the salt-tolerant wheat mutant *RH8706-49* and the salt-sensitive wheat mutant *H8706-34* after treatment with 1% NaCl for 72 h. Qualitative and quantitative differences were found between the two materials for the following five candidate proteins: H^+-transporting two sector ATPase, glutamine synthetase 2 precursor, putative 33 kDa oxygen evolving protein of photosystem II and ribulose-1,5-bisphosphate carboxylase/oxygenase small subunit. All of them are chloroplast proteins, which are likely to play a crucial role in keeping the function of the chloroplast and the whole cells when the plant is under salt stress. Techniques such as two-dimensional electrophoresis and matrix-assisted laser desorption/ionization time of flight mass spectrometry (MALDI-TOF MS) were used to identify proteins affected by salinity in *Triticum durum*. A set of control plants was grown without NaCl addition under the same conditions as the salt-treated plants. The levels of 38 proteins were altered in response to salt stress. In particular, ten of them were down-regulated and 28 were up-regulated (Caruso et al., 2008). The abundance of the proteins related to the carbohydrate and energy metabolism was decreased after NaCl treatment whereas some proteins involved in reactive oxygen species (ROS) scavenging increased. This would indicate that the proteins found in the wheat leaf are involved in several physiological and defence processes. Whilst some are probably part of a general stress response to help plants survive in suboptimal conditions, others may contribute to the reduction of the negative physiological effects of the salt treatment (Caruso et al., 2008).

Zhao et al. (2007) analyzed the expression of all genes in the roots of the salt-tolerant wheat mutant *RH8706-49* by using gene chip technology. The differential expression profile of 61215 genes was gained. The clustering analysis of the hybridization data showed that the gene expression changed greatly in the root under salt stress. Both salt-induced and salt-inhibited genes were found, revealing that the mechanism of salt tolerance in wheat is very complex genetically and it is the result of a lot of gene expression in line.

The expression of the genes encoding the group 3 LEA proteins is thought to be related to stress tolerance in young seedlings. Yu et al. (2004) cloned a novel gene named as *TaLEA 3* from wheat and the results showed that *TaLEA 3* was mainly located in cytoplasm and was induced by high salt, low temperature and exogenous (abscisic acid) ABA treatments. Han et al. (2006) cloned the sequence of *glutamine synthase 2* (*GS2*) from the salt-tolerant wheat mutant *RH8706-49*. It was found that *GS2* was expressed in both mutants, i.e. the salt-tolerant *H8706-49* and the salt-sensitive *H8706-34*, which was the offspring propagated by one-seed mutant. However, the *GS2* expression was significantly higher in the former than in the later, indicating that *GS2* is a kind of salt inhibitory gene and may be regulated at the transcription level.

Ge et al. (2007) cloned the full-length cDNA sequence of the *TaSTK* gene from the salt-tolerant wheat mutant *RH8706-49* using the RACE (rapid amplification of cDNA ends) method. Northern blotting results showed *TaSTK* was a salt-induced gene in wheat. Under salt stress, the expression of *TaSTK* was strongly promoted in the salt-tolerant line *RH8706-49* compared to that in the salt-sensitive line *H8706-34*. But the hybridization signals were all very weak, indicating the *TaSTK* belongs to the low expression genes in the wheat seedling tissue. Mi et al. (2006) obtained a full-length cDNA of a vacuolar ATPase C subunit (*TaVHA-C*) from a wheat line 98–160 via cDNA-AFLP, which has an excellent salt tolerance controlled by major genes. Northern blot analysis showed that both salt and drought stress had a similar effect on the expression of *TaVHA-C*. This gene could be useful to improve plant stress tolerance.

8. MARKER-ASSISTED SELECTION (MAS) FOR SALINITY TOLERANCE

Direct selection of salt-tolerant genotypes under field conditions is hindered by the significant influence that environmental factors have on the plants' response to salinity. There is also evidence supporting the notion that salt tolerance is a complex trait involving the function of many genes. Advances in molecular biology techniques have enabled the development of DNA markers that can be used to identify QTLs. The use of QTLs has improved the efficiency of selection, in particular for those traits that are controlled by several genes and are highly influenced by environmental factors. QTLs and marker-assisted selection (MAS) provide several advantages over direct phenotypic screening, particularly because the PCR-based methodologies used to detect the markers reduce the time needed to screen individuals and reduce the impact of environmental effects on the trait under study. There is considerable evidence to support the view that salt tolerance and its sub-traits are determined by multiple QTLs and that both additive and dominance effects are important in the inheritance of many of the traits associated with salt tolerance (Foolad, 2004; Flowers, 2004).

Salt-related QTLs have been mapped in tomato (Foolad, 2001; Foolad and Jones, 1993), barley (Yoshiro and Kazuyoshi, 1997), soybean (Lee et al., 2004), arabidopsis (Quesada et al., 2002) and rice (Gong et al., 1998; Gu et al., 2000; Lin et al., 1998; 2004; Zhang et al., 1995). Hollington (1998) reported that wheat productivity under saline conditions was largely associated with genes affecting flowering time and ion accumulation and that many QTL effects were around the vernalization response gene *Vrn1* on chromosome 5A. Traits regulating productivity under stress conditions can be identified by comparative QTL analysis and measures of crop productivity such as biomass production or yield. The coincidence of major QTLs for two traits such as leaf ion content and grain yield would indicate their likely association and the identification of alleles of markers associated with major QTLs for improved salt tolerance would allow the techniques to be used to improve the efficiency of selecting for better salt tolerance.

In comparisons between hexaploid, tetraploid and diploid types, it was suggested that bread wheat cultivars have a low rate of Na^+ accumulation and enhanced K^+/Na^+ discrimination, a character located on the long arm of the chromosome 4D (Gorham et al., 1987). This character is controlled by a single locus (*Kna1*) and has been linked to molecular markers on the distal third of the chromosome 4DL (Dubcovsky et al., 1996). *Kna1* is associated with a higher leaf K^+/Na^+ ratio and was attributed to providing bread wheat with its superior salinity tolerance over tetraploid wheats (Gorham et al., 1987). The gene or genes associated with this locus have not been identified. According to Byrt et al. (2007), it is likely an *HKT1;5* gene. A number of attempts were made to map the QTL controlling sodium exclusion, but only one QTL was successfully located on the chromosome 2AL (Munns et al., 2002; Lindsay et al., 2004). In a study of grain yield under salt stress, Quarrie et al. (2005) identified two candidate QTLs, mapping to probable homoeologous regions in the proximal parts of chromosomes 5B and 5D.

The identification of salt tolerance at both the germination and seedling stages is particularly important (Yoshiro and Kazuyoshi, 1997). Ma et al. (2007) used a population of 114 recombinant inbred lines derived from the cross Opata85 × W7984, to find the QTLs related to the response of wheat to salt stress. Forty-seven QTLs were identified mapping all wheat chromosomes except 1B, 1D, 4B, 5D and 7D. Of these QTLs, 10 were effective during the germination stage, and 37 at the seedling stage. The QTLs in the intervals *Xglk683–Xcdo460* (chromosome 3AS) and *Xfbb168–Xbcd147* (chromosome 3BL) were effective at both the germination and seedling stages.

Two dominant genes, which were named *Nax1* and *Nax2* (for Na^+ exclusion) with a major effect for exceptionally low rates of Na^+ accumulation in the leaves, have been identified (Munns et al., 2003). Both genes appear to derive from a *Triticum monococcum* accession, and to be absent in modern tetraploid and hexaploid wheat (James et al., 2006). *Nax1* was mapped to the long arm of chromosome 2A (Lindsay et al., 2004) and one very tightly linked marker, *gwm312*, is being used routinely to select low Na^+ progeny in the durum wheat breeding program. *Nax2* has been mapped (Byrt et al., 2007) and a tightly linked marker is being used for the selection of lines containing *Nax2*. The *Nax* genes have also been transferred into hexaploid wheat cultivars by inter-specific crosses, and the progeny has been selected using the markers (Munns and Richards, 2007). Wu et al. (2006) found that the green fluorescence of the fused protein TaGSK-GFP was distributed in the cytoplasm of transgenic plant's cells and located mainly in the root meristem and stelar sheath, where new roots would develop. The gene *TaGSK1* that was localized in the cytoplasm could improve cell proliferation, development, anti-osmotic and anti-stress ability of the transgenic plants. Li et al. (2006) using Chinese Spring nulli-tetrasomic lines located the gene *TaGSK1* on the short arm of the homoeologous chromosome group 1 in *Triticum aestivum* L. These results could be of benefit to the research on heredity and mechanisms of salt tolerance in wheat.

Wu et al. (2007) mapping QTLs for salt tolerance in the F_2 and F_2/F_3 populations from the cross between the high-yielding variety Taikong 6 and the salt-tolerant variety Dekang 961 identified six QTLs as follows: One QTL for the germination rate was detected on chromosome 5D, two QTLs for the seedling height on chromosome 5D and 5A, two QTLs for the root length on chromosome 5B, and one QTL for fresh weight on chromosome 5D. These results would indicate that the genes controlling wheat salt tolerance at the germination stage may be predominantly located on the homoeologous chromosome group 5. Weng and Chen (2002) identified a major gene controlling the salt-tolerant character in the salt-tolerant variety Chadianhong and cloned a 591 bp marker *OPZ09-591*. This work laid the foundation for further studies on fine cloning and mapping of salt-tolerant genes and MAS for salt tolerance breeding of wheat. Suo et al. (2001) using 280 primers for RAPD analysis of Jimai 24 and its salt-tolerant progeny 8901-17 identified one RAPD marker pronQ4 closely linked to the salt-tolerant mutation.

The facultatively halophytic species *Lophopyrum elongatum* that is closely related to wheat and its amphiploids show better salinity tolerance when they are gradually exposed to salt stress than when they are suddenly stressed. *Lophopyrum elongatum* has greater tolerance to both forms of salt stress than wheat and its genome partially confers this tolerance to its amphiploids. Chromosomal control of the tolerance of both sudden and gradual salt stress in the disomic and ditelosomic addition lines and the disomic substitution lines of *Lophopyrum elongatum* and wheat were studied by Zhong and Dvorak (1995). Wheat chromosomes in the homoeologous groups 1, 3, and 7 and those in the homoeologous groups 1, 4, and 6 were shown to enhance the tolerance of suddenly and gradually imposed salt stress, respectively. Liu et al. (2001) identified a wheat-*Leymus* hybrid line 98–160 with remarkable tolerance to salt stress and found that SSR markers *WMS67* and *WMS213* mapped on 5BL were linked to salt tolerance, which might be controlled by major gene(s).

Wang et al. (2004) mapped salt-tolerant genes in a F_2 wheat population derived from the hybrids of the salt-tolerant mutant *RH8706-49* and the salt-sensitive mutant *H8706-34* by using 246 pairs of micro-satellite primers. Linkage analysis showed that marker *Xgwm299* located on the 3BL was linked to relative salt tolerance. Shan et al. (2006) analyzed a F_2 population of hybridized combinations between a salt-tolerant asymmetric somatic hybrid wheat variety Shanrong No.3 and the salt-sensitive variety Jinan 177. They found that the salt tolerance of Shanrong No.3 was likely controlled by a major gene and the wheat-specific SSR marker *Xgwm304* located on the chromosome 5AS was linked to its salt-tolerant locus. These results were revised later by Chen (2007), who reported that the major gene was located on the wheat chromosome 5AS, between the markers *Xgwm304* and *Xgwm666*.

9. IMPROVEMENT OF SALINITY TOLERANCE BY TRANSFORMATION

Recent advances in the knowledge of gene composition, expression and regulation and signal transformation have enabled scientists to better understand the stress-tolerance mechanisms of plants at the molecular level. The methods of transferring the exogenous target genes related to salt tolerance into wheat via gene engineering with the overall aim at improving the salt tolerance of wheat are intensively receiving more and more attention by a vast number of breeders.

Total genomic DNA from *Agropyron elongatum* ($2n = 70$) was introduced into winter wheat through the pollen tube channels. Variations in plant height, plant type, spike type, disease resistance and stress tolerance were observed. A new variety, Jinan 18, which has a higher saline and drought stress tolerance than the control cv. Lumai No.10, has been developed and released (Huang et al., 2000).

Xue et al. (2004) successfully generated a transgenic wheat by expressing a vacuolar Na^+/H^+ antiporter gene *AtNHX1* from *Arabidopsis thaliana* to enhance plants' capacity to reduce cytosolic Na^+ by sequestering Na^+ in the vacuole. In field plots, grain yield of the best T_3 transgenic wheat line grown in moderately saline soil (EC_e=10.6 dS m^{-1}) was 50% of that under non-saline conditions, whereas the yield of the non-transformed control was only 34% of that under non-saline conditions. At a higher salinity level (EC_e=13.7 dS m^{-1}), grain yield of the best transgenic line was 18% of its non-saline control, while yield of the non-transformed control was still lower at 11%. These gains in salt tolerance appear modest and convincing (Xue et al., 2004).

Munns and Richards (2007) reported that the ornithine aminotransferase (OAT) gene has been transferred into two Australian commercial wheats by Grain Biotech Australia. Glasshouse studies showed that two lines had 2–5 fold higher yields and 2-fold higher levels of free proline than the control varieties, when the plants were salt-stressed. In a field assessment of the transgenic lines on a salinity gradient, it was found that they had significantly higher yields than the commercial parents at the high salinity levels, although not as great as when grown in the non-saline soil.

Liu et al. (2006) introduced the antisense phospholipase Dγ (PLDγ) gene into the wheat variety Fuxiwumai by means of ear stem injection. Ten transformed individual plants were obtained from the T_1 generation with PCR and Southern blotting test. From these, six plants had increased tolerance to cold injury and two plants showed improved salt tolerance in the MS medium containing NaCl.

Ji et al. (2002) successfully introduced the *HVA1* gene, one of the members of late embryogenesis abundant (LEA) protein family, into wheat by the deflating pollen tube method, in order to produce salt-tolerant wheat lines. This gene plays an important role in the plant survival under stresses such as salt and drought.

A trehalose-6-phosphate synthase (TPS) gene from *Saccharomyces cerevisiae* related to plant drought and salt tolerance was transferred into the common wheat variety CB9945 by biolistic bombardment. Fifteen transgenic lines with improved stress tolerance to some extent were obtained under simulated drought and salt stress conditions. Some of the lines have been tested under controlled field conditions (Du et al., 2007).

The *Tagsk1* (*Triticum aestivum* L. glycogen synthase kinase 1) gene cloned from the genome of the wheat salt-tolerant mutant *RH8706-49* was introduced into the induced callus from mature embryos of the salt-sensitive wheat *H8706-34* and the variety Chinese Spring by particle bombardment. The transformed callus stressed by Kanamycin and 0.5% NaCl showed a higher ability of salt tolerance and was able to differentiate roots and buds on the medium containing 0.5% NaCl (Xu et al., 2006).

Guo et al. (2000) reported that the betaine aldehyde dehydrogenase (*BADH*) cDNA cloned from *Atriplex hortensis* L. was transferred into wheat. Under salt-stressed conditions, transgenic plants showed 100%–300% higher *BADH* activity than the control. Further identification showed that the transgenic wheat lines had several advantages over their donor plants, for instance, improved germination ability, seedling growth and root development under drought stress condition as well as enhanced plasma membrane protection of the excised leaf and lower transpiration rate under field conditions (Zhang et al., 2003).

It has been found that under abiotic stress conditions such as drought, high-salinity and low-temperature, the transcription factor of DREB (dehydration responsive element binding proteins) improved efficiently plants stress resistance by regulating the super expressions of a series of genes related to stressful environment conditions. Cloning and transforming some main transcription factors into plants is becoming an effective approach to enhancing crop stress resistance. Based on *in vitro* culture methods, a new efficient regeneration system for wheat genetic transformation without genotype dependence has been explored and set up (Liu et al., 2001). By use of this system, DREB genes cloned from *Arabidopsis* were transferred into wheat and transgenic plants were obtained (Liu et al., 2003b). The *GmDREB* gene encoding a stress-inducible transcription factor was also cloned and transformed into wheat variety Lumai 22 by bombardment. Some promising transgenic lines containing the *GmDREB* gene with good drought or salt tolerance as well as similar agronomical characters to the variety Lumai 22 were obtained. The improved drought and salt tolerance of the T_1 transgenic lines was demonstrated in comparison to the wild type (Gao et al., 2005).

REFERENCES

Barakat, M. N., & Abdel-Latif, T. H. (1996). *In vitro* selection of wheat callus tolerant to high levels of salt and plant regeneration. *Euphytica*, *91*, 127-140.

Blum, A., & Ebercon, A. (1981). Cell membrane stability as a measure of drought and heat tolerance in wheat. *Crop Sci*, *21*, 43-47.

Blum, A. (1988). Plant Breeding for Stress Environments, CRC Press, Boca Raton, Florida, 233.

Byrt, C., et al. (2007). *HKT1;5*-like cation transporters linked to Na^+ exclusion loci in wheat, *Nax2* and *Kna1*. *Plant Physiol*, *143*, 1918-1928.

Caruso, G., et al. (2008). Identification of changes in *Triticum durum* L. leaf proteome in response to salt stress by two-dimensional electrophoresis and MALDI-TOF mass spectrometry, *Anal. Bioanal. Chem*, *391*, 381-390.

Chen, F. (2007). *The major salt-relative gene located by SSR markers in salt tolerance introgression F_2 population of Shanrong No. 3 with Jinan 177*, Master's Thesis, Shandong University.

Chen, S. Y., et al. (2004). Introgression of salt tolerance from somatic hybrids between common wheat and *Thinopyrum ponticum*. *Plant Sci*, *167*, 773-779.

Colmer, T. D., et al. (2006). Use of wild relatives to improve salt tolerance of wheat. *J. Exp. Bot*, *57*, 1059-1078.

Ding, T. l., et al. (2006). Na^+/K^+ selectivity of leaf sheath in wheat cultivars differing in salt tolerance. *J. Plant Physiol. Mol. Biol*, *32*(1), 123-126.

Dracup, M. (1991). Increasing salt tolerance of plants through cell culture requires greater understanding of tolerance mechanisms. *Aust. J. Plant Physiol*, *18*(1), 1-15.

Du, L. P., et al. (2007). Transgenic wheat plants with trehalose-6-phosphate synthase (TPS) gene and identification of their function. *J. Triticeae Crops*, *27*(3), 369-373.

Dubcovsky, J., et al. (1996). Mapping of the K/Na discrimination locus *Knal* in wheat. *Theor. Appl. Genet*, *92*, 448-454.

El-Hendawy, S. E., et al. (2005). Growth, ion content, gas exchange, and water relations of wheat genotypes differing in salt tolerances. *Aust. J. Agric. Res*, *56*(2), 123-134.

Farooq, S. (2004). Salt tolerance in *Aegilops* species: A success story from research and production to large-scale utilization of salt-tolerant wheat, In Taha, F. K., Ismail, S., & Jaradat, A. (Eds.), *Prospects of Saline Agriculture in the Arabian Peninsula*, Amherst Scientific Publishers, Massachusetts, 121-134.

Farooq, S., & Azam, F. (2005). Salinity tolerance in *Triticeae*. *J. Genet. Plant Breed*, *41*, 252-262.

Farooq, S., & Azam, F. (2006). The use of cell membrane stability (CMS) technique to screen for salt-tolerant wheat varieties. *J. Plant Physiol*, *163(6)*, 629-637.

Farooq, S., et al. (1992a). Intergeneric hybridization for wheat improvement. IV. Expression of salt tolerance gene(s). of *Aegilops cylindrica* in hybrids with hexaploid wheat. *Cereal Res. Commun*, *20*, 111- 118.

Farooq, S., et al. (1992b). Intergeneric hybridization for wheat improvement. VI. Production of salt-tolerant germplasm through crossing wheat (*Triticum aestivum* L.) with *Aegilops cylindrica* and its significance in practical agriculture. *J. Genet. Breed*, *46*, 125-132.

Farooq, S., et al. (1995). Production of salt-tolerant wheat germplasm through crossing cultivated wheat (*Triticum aestivum* L.) with *Aegilops cylindrica*. Ⅱ. Field evaluation of salt-tolerant germplasm. *Cereal Res. Commun*, *23*, 275-282.

Flowers, T. J. (2004). Improving crop salt tolerance. *J. Exp. Bot*, *55*, 307-319.

Flowers, T. J., et al. (1977). The mechanism of salt tolerance in halophytes. *Annu. Rev. Plant Physiol*, *28*, 89-121.

Foolad, M. R. (2004). Recent advances in genetics of salt tolerance in tomato. *Plant Cell, Tissue Organ Cult*, *76*, 101-119.

Foolad, M. R., & Jones R.A. (1993). Mapping salt tolerance genes in tomato (*Lycopersicon esculentum*) using trait-based marker analysis. *Theor. Appl. Genet*, *87*, 184-192.

Foolad, M. R. (2001). Identification and validation of QTLs for salt tolerance during vegetative growth in tomato by selective genotyping. *Genome*, *44*, 444-454.

Forster, B. P. (1994). Cytogenetic manipulations in the *Triticeae*, In Yeo,A. R., & Flowers, T. J. (Eds.), *Monographs on Theoretical and Applied Genetics*, Vol. 21, Springer-Verlag, Berlin Heidelberg.

Forster, B. P. (2001). Mutation genetics of salt tolerance in barley: an assessment of Golden Promise and other semi-dwarf mutants. *Euphytica*, *120*, 317-328.

Forster, B. P., & Miller, T. E. (1985). A 5B deficient hybrid between *Triticum aestivum* L. and *Agropyron junceum*. *Cereal Res. Commun*, *13*, 93-95.

Forster, B. P., et al. (1988). The use of genetic stocks in understanding and improving the salt tolerance in wheat, In Joma M. L., & Slootmaker, L. A. J. (Eds.), *Cereals Breeding Related to Integrated Cereal Production*, Wageningen, PUDOC.

Forster, B. P., et al. (1994). Variation in the leaf sodium content of *Hordeum vulgare* (barley) cultivar Maythorpe and its derived mutant cv. Golden Promise. *Heredity*, *73*, 249-253.

Gao, S. Q., et al. (2005). Improvement of wheat drought and salt tolerance by expression of a stress-inducible transcription factor GmDREB of soybean (*Glycine max*). *Chin. Sci. Bull*, *50*(23), 2714-2723.

Ge, R. C., et al. (2007). Cloning of salt tolerance related gene *TaSTK* in *Triticum aestivum* L. *Acta Agron. Sin*, *33*(5), 857-860.

Gong, J. M., et al. (1998). Mapping QTLs related salt tolerance in rice. *Chin. Sci. Bull*, *43*, 1847-1850.

Gorham, J., et al. (1997). Genetic analysis and physiology of a trait for enhanced K^+/Na^+ discrimination in wheat. *New Phytol*, *137*, 109-116.

Gorham, J., et al. (1986). Salt tolerance in the *Triticeae*: solute accumulation and distribution in an amphidiploid derived from *Triticum aestivum* L. cv. Chinese Spring and *Thinopyrum bessarabicum*. *J. Exp. Bot*, *37*, 1435-1449.

Gorham, J., et al. (1987). Chromosomal location of a K/Na discrimination character in the D genome of wheat. *Theor. Appl. Genet*, *74*, 484-488.

Greenway, H., & Munns, R. (1980). Mechanisms of salt tolerance in non-halophytes. *Annu. Rev. Plant Physiol*, *31*, 149-190.

Gu, X. Y., et al. (2000). Preliminary detection of quantitative trait loci for salt tolerance in rice. *Chin. J. Rice Sci*, *14*, 65-70.

Gulick, P., & Dvorak, J. (1987). Gene induction and repression by salt treatment in roots of the salinity-sensitive Chinese Spring wheat and the salinity-tolerant Chinese Spring × *Elytrigia elongatum* amphiploid. *Proc. Natl. Acad. Sci. USA*, *84*, 99-103.

Guo, B. H., et al.(2000). Transformation of wheat with a gene encoding for betaine aldehyde dehydrogenase (*BADH*). *Acta Bot. Sin*, *42*(3), 279-283.

Guo, B. S., et al. (2001). Identification and analysis of salt tolerance in Tibetan wheat. *J. Plant Genet. Resour*, *2*(2), 36-39.

Guo, F. Q., et al. (1997). Mutation, selection and comparison of several saline-tolerant wheat strains. *Acta Agric. Nucl. Sin*, *11*(1), 1-8.

Han, N., et al. (2006). Cloning and analysis of glutamine synthase 2 (GS2) precursor in *Triticum aestivum* L. *Acta Agron. Sin*, *32(*11), 1756-1758.

Hasegawa, P. M., et al. (2000). Plant cellular and molecular responses to high salinity. *Annu. Rev. Plant Physiol. Plant Mol. Biol*, *51*, *463-499*.

Hollington, P. A. (1998). Technological breakthroughs in screening/breeding wheat varieties for salt tolerance. *Proceedings of the national conference on salinity management in agriculture*, 2-5.

Hossain, Z., et al. (2006). Isolation of a NaCl-tolerant mutant of *Chrysanthemum morifolium* by gamma radiation: *in vitro* mutagenesis and selection by salt stress. *Funct. Plant Biol*, *33*(1), 91-101.

Hou, N., et al. (2004). Genetic effects of *Aegilops crassa* 6x cytoplasm on salt tolerance of common wheat. *J. Triticeae Crops*, *24*(2), 5-10.

Huang, C. Y., et al. (2000). Jinan 18, a new wheat variety with drought resistance and salt tolerance. *China Seed Industry*, *5*, 45.

Huo, C. M., et al. (2004). Proteomic analysis of the salt tolerance mutant of wheat under salt stress. *Acta Genet. Sin*, *31*(12), 1408-1414.

Husain, S., et al. (2003). Effect of sodium exclusion trait on chlorophyll retention and growth of durum wheat in saline soil. *Aust. J. Agric. Res*, *54*, 589-597.

International Atomic Energy Agency (IAEA). (2002). *Novel approaches for improving crop tolerance to salinity and drought*, Report of a group consultant meeting held under TC project INT5144, IAEA, working material, Vienna.

Jain, S. M. (2001). Tissue culture-derived variation in crop improvement. *Euphytica*, *118*, 153-166.

James, R. A., et al. (2006). Physiological characterisation of two genes exclusion in wheat: *Nax1* and *Nax2*. *Plant Physiol*, *142*, 1537-1547.

Ji, J. L., et al. (2002). A study on the transformation of wheat with salt-tolerant gene *HVA1* by method deflating pollen tube. *J. Triticeae Crops*, *22*(2), 10-13.

King, I. P., et al. (1996). An assessment of the salt tolerance of wheat/*Thinopyrum bessarabicum* 5Eb addition and substitution lines. *Plant Breed*, *115*, 77-78.

King, I. P., et al. (1997). *Tritipyrum*, a potential new salt-tolerant cereal. *Plant Breed*, *116*, 127-132.

Kingsbury, R.W., & Epstein, E. (1984). Selection for salt-resistant spring wheat. *Crop Sci*, *24*, 310-315.

Lee, G. J., et al. (2004). A major QTL conditioning salt tolerance in S-100 soybean and descent cultivars. *Theor. Appl. Genet*, *109*, 1610-1619.

Li, Y. Q., et al. (2006). The chromosomal location of *Triticum aestivum* L. glycogen Synthase Kinase Gene (TaGSK). *Acta. Agric. Boreali-Sinica*, *21*(5), 39-41.

Liang, C., et al. (2006). Salt-tolerant physiological characters of wheat variety Dekang 961. *Acta Bot. Boreali-Occident. Sin*, *26*(10), 2075-2082.

Lin, H. X., et al. (2004). QTL for Na and K uptake of the shoot and roots controlling rice salt tolerance. *Theor. Appl. Genet*, *108*, 253-260.

Lin, H. X., et al. (1998). Identification of QTL for salt tolerance in rice via molecular markers. *Chin. J. Rice Sci*, *12*, 72-78.

Lindsay, M. P., et al. (2004). A locus for sodium exclusion (*Nax1*), a trait for salt tolerance mapped in durum wheat. *Funct. Plant Biol*, *31*, 1105-1114.

Liu, F., et al. (2006). Transformation of wheat with anti-phospholipase D γ gene. *Lett. Biotechnol*, *17*(2), 189-191.

Liu, L. X., et al. (2001). Study on receptor system for wheat genetic transformation without genotype dependence. *Acta Agric. Nucl. Sin*, *15*(5), 308-310.

Liu, L. X., et al. (2003a). *Salt tolerance improvement by using mutation techniques and biotechnology in wheat*, Proceedings on Developing Salt-tolerant Crops for Sustainable Food and Feed Production in Saline Lands, 10-19.

Liu, L. X., et al. (2003b). Study on production of transgenic wheat with a stress-inducible transcription factor gene *DREB1A* by microprojectile bombardment. *China Biotechnol*, *23*(11), 53-56.

Liu, L. X., et al. (2006). *Salt tolerance improvement through doubled haploid and nuclear radiation technique in wheat*, Abstracts 11th IAPTC&B Congress Biotechnology and Sustainable Agriculture 2006 and Beyond, 191.

Liu, L. X., et al. (2007). A salt-tolerant mutant wheat cultivar 'H6756'. *Plant Mutation Reports*, *1*(3), 50-51.

Liu, L. X., et al. (2001). Screening salt tolerance germplasms and tagging the tolerance genes using microsatellite (SSR) markers in wheat. *Acta Bot. Sin*, *43*(9), 948-954.

Luo, T. B., & Ren, W. (2001). Breeding and characters of the new line of salt-tolerant and high-yielding wheat"101"in Xinjian. *Agric. Res. in the Arid Areas*, *19*(1), 131-132.

Luo, T. B., et al. (2005). Salt tolerance of early combinations in wheat diallel crossings. *Agric. Res. in the Arid Areas*, *23*(5), 80-84.

Luo, T. B., et al. (2006). Primary agronomic properties and breeding process of new lines of salt-tolerant wheat in Xinjiang. *Agric. Res. in the Arid Areas*, *24*(2), 18-21.

Ma, Y. Q., & Weng, Y. J. (2005). Evaluation for salt tolerance in spring wheat cultivars introduced from abroad. *Acta Agron. Sin*, *31*(1), 58-64.

Mi, C. L., et al. (2006). Isolation of *TaVHA-C*, a gene in wheat related to salt tolerance via cDNA-AFLP. *Sci. Agric. Sin*, *39*(9), 1736-1742.

Mujeeb-Kazi, A., & Diaz de Leon, J. L. (2002). Conventional and alien genetic diversity for salt-tolerant wheats: focus on current status and new germplasm development, In Ahmad, R., & Malik, K. A. (Eds.), *Prospects for Saline Agriculture*, vol. 37, Kluwer Academic Publishers, Dordrecht, 69-82.

Munns, R. (2005). Genes and salt tolerance: bringing them together. *New Phytol*, *167*, 645-663.

Munns, R., & James, R. A. (2003). Screening methods for salinity tolerance: a case study with tetraploid wheat. *Plant Soil*, *253*, 201-218.

Munns, R., & Richards, R. A. (2007). Recent advances in breeding wheat for drought and salt stresses, Chapter 22, In Jenks, M. A., et al. (Eds.), *Advances in Molecular Breeding toward Drought and Salt-Tolerant Crops*, Springer Verlag, Berlin, 565-585.

Munns, R., & Tester, M. (2008). Mechanisms of salinity tolerance. *Annu. Rev. Plant Biol*, *59*, 651-68.

Munns, R., et al. (2002). Avenues for increasing salt tolerance of crop and the role of physiologically-based selection traits. *Plant Soil*, *247*, 93-105.

Noble, C. L. (1983). The potential for breeding salt-tolerant plants. *Proc. Royal Soc.Victoria*, *95*, 133-138.

Noble, C. L., & Rogers, M.E. (1992). Arguments for the use of physiological criteria for improving salt tolerance in crops. *Plant Physiol*, *146*, 99-107.

Noori, S. A. S. (2005). Assessment for salinity tolerance through intergeneric hybridisation:

Triticum durum × *Aegilops speltoides*. *Euphytica*, *146*, 149-155.

Norlyn, J. D. (1980). Breeding salt-tolerant crop plants, In Rains, D.W., et al. (Eds.), *Genetic regulation of osmo-regulation*, Plenum Press, New York, 239-309.

Pang, C. H., & Wang, B. S. (2008). Oxidative stress and salt tolerance in plants, In Lüttge, U., et al. (Eds.), *Progress in Botany 69*, Springer Verlag, Berlin, 231-245.

Premachandra, G. S., et al. (1991). Cell membrane stability and leaf surface wax content as affected by increasing water deficits in maize. *J. Exp. Bot*, *42*, 167-171.

Quarrie, S. A., et al. (2005). A high-density genetic map of hexaploid wheat (*Triticum aestivum* L.) from the cross Chinese Spring × SQ1 and its use to compare QTLs for grain yield across a range of environments. *Theor. Appl. Genet*, *110*, 965-990.

Qureshi, R. H., et al. (1980). Screening wheat (*Triticum aestivum* L.) for salt tolerance. *Pak. J. Agric. Sci*, *17*, 19-25.

Quesada, V., et al. (2002). Genetic architecture of NaCl tolerance in *Arabidopsis*. *Plant Physiol*, *130*, 951-963.

Rahman, M. H., et al. (1995). Selection for salt tolerance *in vitro* using microspore-derived embryos of *Brassica napus* cv Topas, and the characterization of putative tolerant plants. *In Vitro* Cell. Dev. Biol.: *Plant*, *31*, 116-121.

Sayed, J. (1985). Diversity of salt tolerance in a germplasm collection of wheat (*Triticum aestivum* L.). *Theor. Appl. Genet*, *69*, 651-657.

Schachtman, D. P., et al. (1992). The expression of salt tolerance from *Triticum tauschii* in hexaploid wheat. *Theor. Appl. Genet*, *84*, 714-719.

Shan, L., et al. (2006). Screening and localization of SSR markers related to salt tolerance of somatic hybrid wheat Shanrong No.3. *Sci. Agric. Sin*, *9*(2), 225-230.

Shannon, M. C. (1984). Breeding, selection and the genetics of salt tolerance, In Staples, R. C., & Toeniessen, G. A. (Eds.), *Salinity Tolerance in Plants: Strategies for Crop Improvement*, John Wiley and Co., New York, USA, 231-254.

Sharma, S. K., & Goyal S. S. (2003). Progress in plant salinity resistance research: need for an integrative paradigm. *J. Crop Prod*, *7*(1/2), 387-407.

Shen, Y. Z., et al. (1993). A study of the salt-resistant variation of inducing matured embryo's callus and regenerated plants in wheat. *Acta Genet. Sin*, *20*(3), 283-291.

Shen, Y. Z., et al. (1997). A study of salt-resistant variations induced in anther calli and regenerated plants in wheat. *Hereditas*, *19*(6), 7-11.

Singh, K. N., & Chatrath, R. (1992). Genetic variability in grain yield and its component characters and their associations under salt stress conditions in tissue culture lines of bread wheat (*Triticum aestivum* L. em. Thell). *Wheat Information Service*, *75*, 46-53.

Suo, G. L., et al. (2001). Identification of the molecular markers linked to the salt-resistance locus in the wheat using RAPD-BSA technique. *Acta Bot. Sin*, *43*(6), 598- 602.

Szarejko, I., & Forster, B. P. (2007). Doubled haploidy and induced mutation, *Euphytica*, *158*, 359-370.

Tal, M. (1994). *In vitro* selection for salt tolerance in crop plants: theoretical and practical considerations. *In Vitro Cell. Dev. Biol.: Plant*, *30*, 175-180.

Tester, M., & Davenport, R. (2003). Na^+ tolerance and Na^+ transport in higher plants. *Ann. Bot. 91*, 503-527.

Trethowan, R. M., & Mujeeb-Kazi, A. (2008). Novel germplasm resources for improving environmental stress tolerance of hexaploid wheat. *Crop Sci*, *48*, 1255-1265.

Tuna, A. L., et al. (2008). Silicon improves salinity tolerance in wheat plants. *Environ. Exp. Bot*, *62*(1), 10-16.

Wang, H. Y., et al. (2004). Mapping of relative salt tolerance gene in wheat salt-tolerant mutant by using microsatellite marker. *Acta Agron. Sin*, *30*(7), 697-699.

Wang, R. R. C., et al. (2003). Registration of W4909 and W4910 bread wheat germplasm lines with high salinity tolerance. *Crop Sci*, *43*, 746.

Weng, Y. J., & Chen, D. M. (2002). Molecular markers and its clone for salt tolerance gene. *Acta Genet. Sin*, *29*(4), 343-349.

Wu, L. Z., et al. (2006). Localization and function analysis of wheat glycogen synthesis kinase (*TaGSK1*). *Sci. Agric. Sin*, *39*(4), 842-847.

Wu, Y. Q., et al. (2007). Mapping QTL for salt-tolerant traits in wheat. *J. Nucl. Agric. Sci*, *21*(6), 545.

Xiao, H. L., & Zhao, S. X. (1989). The selection of NaCl-tolerant mutants by tissue culture in wheat. *Acta Agric. Nucl. Sin*, *3*(2), 85-90.

Xu, T., et al. (2006). Introduce *Tagsk1* into salt-sensitive callus to improve the capacity of salt tolerance by microparticle bombardment. *Chin. J. Biotechnol*, *22*(2), 211-214.

Xue, Z. Y., et al. (2004). Enhanced salt tolerance of transgenic wheat (*Triticum aestivum* L.) expressing a vacuolar Na^+/H^+ antiporter gene with improved grain yields in saline soils in the field and a reduced level of leaf Na^+. *Plant Sci*, *167*, 849-859.

Yeo, A. R. (1983). Flowers T. J., Varietal differences in the toxicity of the sodium ions in rice leaves. *Physiol. Plant*, *59*, 189-195.

Yeo, A. R., & Flowers T. J. (1986). Salinity resistance in rice (*Oryza sativa* L.) and a pyramiding approach to breeding varieties for saline soils. *Aust. J. Plant Physiol*, *13*, 161-173.

Yoshiro, M., & Kazuyoshi, T. (1997). Mapping quantitative trait loci for salt tolerance at germination stage and the seedling stage in barley, (*Hordeum vulgare* L.). *Euphytica*, *94*, 263-

272.

Yu, J. N., et al. (2004). Cloning of a novel stress tolerant gene-*TaLEA3* from wheat and the functional analysis in yeast. *Chin. J. Biotechnol*, *20*(6), 832-838.

Zhang, G. Y., et al. (1995). RFLP tagging of a salt tolerance gene in rice. *Plant Sci*, *110*, 227-334.

Zhang, Y. M., et al. (2003). Salt and drought stress tolerance in transgenic wheat expressing betaine aldehyde dehydrogenase gene. *Acta Agric. Boreali-Sinica*, *18*(1), 29-32.

Zhao, B. C., et al. (2007). Study on the expression profile of salt tolerance mutant under salt-stress in wheat using gene microarray. *Sci. Agric. Sin*, *40*(10), 2355-2360.

Zhao, R. T., et al. (1994). A new approach of screening salt tolerance variants by anther culture to cultivate salt tolerance wheat variety. *Acta Agric. Boreali-Sinica*, *9*(1), 34-38.

Zhao, R. T., et al. (1995). Studies on the application of anther culture in salt tolerance breeding in wheat (*Triticum aestivum* L.). *Acta Agron. Sin*, *21*(2), 230-234.

Zhao, S. L., & Dou, Y. L. (1998). A review of identification indicators for wheat salt tolerance. *J. Northwest China Coll. Agric*, *26*(6), 80-85.

Zhao, S. S., et al. (2000). Breeding and selection of drought resistant and salt-tolerant wheat variety Cang6001. *Acta Agric. Boreali-Sinica*, *15*(supplement), 113-117.

Zheng, Q. C., et al. (1994). A study on salt stress induced proteins in the salt-tolerant mutant in wheat. *J. Nucl. Agric. Sci*, *15*(3), 101-104.

Zheng, Q. C., et al. (1996). Screening of salt-tolerant wheat via an unicell and haploid system. *J. Nucl. Agric. Sci*, *17*(6), 253-255.

Zheng, Y. H., et al. (2008). Higher salinity tolerance cultivars of winter wheat relieved senescence at reproductive stage. *Environ. Exp. Bot*, *62*, 129-138.

Zhong, G. Y., & Dvorak, J. (1995). Chromosomal control of the tolerance of gradually and suddenly imposed salt stress in *Lophopyrum elongatum* and wheat, *Triticum aestivum* L., genomes. *Theor. Appl. Genet*, *90*, 229-236.

Zhu, J. K. (2002). Salt and drought stress signal transduction in plants. *Annu. Rev. Plant Biol*, *53*, 247-273.

Zhu, Z. H., et al. (1996). Identification and studies of salt tolerance in wheat, In Zhang Q. S., & Du Z. H. (Eds.), *Advances in Wheat Breeding Research in China*, Chin. Agric. Press, Beijing 287-294.

Zulfiqar, A. (2004). *Genetic basis of salt tolerance in wheat*, PhD thesis, University of Agriculture, Faisalabad.

CHAPTER 5.

PROGRESS ON THE PHYSIOLOGY AND GENETIC RESEARCH FOR SALT TOLERANCE IN RICE

Dongling Qi[1], Guizhen Guo[3], Myung-chul Lee[4], Guilan Cao[1], Junguo Zhang[3], Qingyang Zhou[2], Sanyuan Zhang[3], Seok-cheol Suh[4] and Longzhi Han[1*]

[1]Key Laboratory of Crop Germplasm Resources and Biotechnology, Ministry of Agriculture and Rural Affairs of the People's Republic of China/National Key Facility for Crop Gene Resources and Genetic Improvement /Institute of Crop Sciences, Chinese Academy of Agricultural Sciences, Beijing 100081, China;

[2]Institute of Rubber Research, Chinese Academy of Tropical Agricultural Sciences, Danzhou 571737, China;

[3]Institute of Rice Research, Jilin Academy of Agricultural Sciences, Gongzhuling 136100, China;

[4]National Institute of Agricultural Biotechnology, RDA, Suwon 441-707, Korea

[*]Correspondence: lzhan58@yahoo.com.cn

1. INTRODUCTION

Salinization causes worldwide soil and water degradation processes, which represents a serious threat to agricultural production. There are major abiotic stresses affecting normal growth and yields of cultivated plant crops. It is reported that the crop output reduction due to damages from drought and salinity stresses is above 40%. The global extent of salt-affected soils is about 9×10^8 ha, which accounts for 20% of total land area (Munns, 2002).

In China the estimated area of salt-affected soils is about 2×10^7 ha. With the rapid growth of the Chinese population and the increased urbanization and industrialization, the available land *per capita* is continuously decreasing while land degradation problems such as salinity are on the rise. Moreover, in recent years the availability of freshwater resources for agriculture is becoming scarcer. As a result, agricultural production is confronted with serious challenges, especially in arid and semiarid environments and coastal land areas affected by salinity. Therefore, the amelioration and sustainable utilization of these salt-affected soils for crop production is an important problem requiring urgent solutions.

Generally speaking, the salt tolerance of rice is defined as its capability of endurance and resistance under saline-alkaline stress conditions. Rice is classified as moderately sensitive to salinity stress. However, salt-affected soils cause serious effects on the rice plants leading to reduced grain production. An in-depth study of the rice salt tolerance is critical to improve the yield potential of rice in salt-affected soils, for enlarging the rice cultivated area, and ensuring food security in the rice-growing areas. These impacts are also relevant to improve peoples' welfare and environmental sustainability. This chapter describes recent progress achieved in the genetic improvement of rice for salt tolerance, with particular reference to rice growth and development and agronomic traits affected by salinity stress, physiological mechanisms and genetics of rice salinity tolerance, molecular studies of salt tolerance in rice and gene cloning and transformation of rice for salt tolerance.

2. GENETIC DIFFERENCES OF SALT TOLERANCE IN RICE

2.1 Salinity tolerance in rice

Rice is a plant native to wetlands and is moderately sensitive to salinity stress (Wang et al., 1986). Root length, root number and seedling length of rice were restrained at 200 mmol L^{-1} NaCl concentration at the budding period, and rice seedlings at 4-leaf-stage died quickly at 283 mmol L^{-1} Na_2CO_3 concentration (Cheng et al., 1994). Under salinity stress, rice tillering capability per plant decreased (Liang et al., 2004). Cheng et al. (1995) reported a soil pH 8.6 as the critical value for screening rice for alkalinity tolerance at mid-tillering and booting stages. Salinity and alkalinity tolerances are not correlated within a rice variety (Cheng et al., 1996).

2.2 Differences of salinity tolerance among rice varieties

The haloduric limit is different among japonica rice varieties (Zhang et al., 2006). In addition, salt tolerance differences can be found in different varieties, which could be closely correlated with the historic origin, evolution and genetic background of rice.

2.3 Differences of salinity tolerance among development stages

Guo et al. (2004) reported that the salinity tolerance at germination and seedling stages was different for the same rice variety. Akbar et al. (1985) pointed out that salt tolerance in rice was strong at early growth stages and the rice crop became more sensitive to salt stress at flowering and seed-filling stages. These salinity tolerance differences among growth stages could be closely correlated with the genetic characteristics and physiological/chemical response of the varieties.

2.3.1 Germination and emergence

Salinity stress can delay germination and growth of rice seedlings (Cssri and Girdhar, 1988). Salinity stress at seed germination stage in rice will lead to high germination variability, reduced germination vigour and decreased germination rate. Up to now, the mechanisms of tolerance to salinity stress at rice seed germination are still poorly understood, whereas the salt stress mainly constrained water absorption during rice seed development. Yan et al. (1995) reported that salinity stress first destroyed membranes during the imbibition of rice seeds and then some physiological processes were affected.

2.3.2 Seedling growth

Salinity stress in rice seedlings leads to leaf curling or death, restrained leaf expansion and limited foliar and root growth. Also both root length and seedling height decreased at the seedling stage (Khan et al., 1997; Shan et al., 2006).

2.3.3 Plant development and maturity

Salinity stress is one of the main factors affecting unfavourably plant photosynthesis. Liang et al. (2004) indicated that the heading stage was delayed or tillering was impaired. The heading stage occurred later in the salt-sensitive early rice varieties compared to the moderately salt-tolerant late rice varieties. Under salinity stress, the plant height was reduced, and the number of tillers and green leaves decreased (Cheng et al., 1996). The rice leaves are mostly affected at tillering stage. The production index including the length of internodes and flowers seriously decline; the effective panicle, 1,000-grain weight and the number of tillers decrease (Lee et al., 2002).

Rice seedlings sensitive to salt stress are severely damaged by excess ions. These ion toxicities affect photosynthesis by disturbing trans-stomatal movement and reducing CO_2 uptake (Wang et al., 2002). Brugnoli and Lauteri (1991) postulated that the effect of salinity on the CO_2 assimilation rate was mostly due to the reduction of stomatal conductance. Stomatal closure in the salt-sensitive rice varieties was the main reason for this (Li et al., 2006). Rice panicle formation is seriously hampered and seed set rate is strongly reduced by salinity (Khan and Abdullah, 2003). Salinity generally reduced plant vigour, produced stunted plant growth, reduced straw and grain weight, and induced panicle sterility with large variations between lines (Shereen et al., 2002; Zhang et al., 2006).

2.4 Differences of salinity tolerance among different parts in rice plants

Plant height and length of the internodes are not sensitive to salt stress at rice maturity stage, but some developmental tissues such as young leaf, radicle and inflorescences are more sensitive to salt stress. Furthermore, the above-ground parts of young rice seedlings are more sensitive to salt stress than the underground parts (Han et al., 1998). From the above, plant height and length of internodes are not suitable indicators of level of salt tolerance at rice maturity but tillering capability could be used as the criterion of salinity tolerance of rice varieties (Liang et al., 2004). Salinity tolerance rating at the seedling stage can be made using relative plant height and relative root dry weight, which are criteria visually different among rice varieties at the same temperature and salt stress level.

2.5 Effect of salinity stress on development and agronomic traits of rice

Excessive Na^+ accumulation in rice paddies antagonizes the absorption of major nutrient elements and microelements. Such antagonism processes induce crop physiological disorders and metabolism interruptions.

3. PHYSIOLOGICAL MECHANISMS FOR SALT TOLERANCE IN RICE

Plants grown in salt-affected soils are confronted with two kinds of abiotic stress, namely osmotic and specific ion stress (Shu et al., 2007). Plants can grow normally in these conditions if they are able to overcome these stresses (Zhao, 2002). Salinity tolerance of higher plants includes both mechanisms of ion uptake and ion exclusion. Restriction of ions to entering into above-ground parts of plants is an important indicator of the capability of salinity tolerance (Greenway and Munns, 1980).

3.1 Adjustments in uptake and transport of mineral ions

Under NaCl stress, the content of Na^+ in the stele parenchyma of the salt-tolerant genotypes is relatively higher than in the other parts of the roots while in the salt-sensitive genotypes the Na^+ content is evenly distributed in the roots. This implies that the stele parenchyma might be the site controlling Na^+ translocation from the root to the shoot (Zheng et al., 1996). A short-term (up to 24 h) study was conducted to study the effect of NaCl stress on the ion transport and nutritional and water status of rice. The results showed that the NaCl stress markedly affected the ion composition of the xylem sap of rice, and a rapid and remarkable increase in the concentrations of K^+, Ca^{2+} and Mg^{2+}. In rice, the Na^+ transport is restricted whilst the K^+ transport rapidly increases.

This results in a high ability of osmotic adjustment and an adequate nutritional status (Baba and Fujiya, 2003). The concentration and uptake of ions were studied in rice seedlings under different salinity levels. The root dry matter was significantly reduced and the K^+ concentration decreased significantly in shoots at all levels. This may indicate that selective ion absorption would be the principal salt tolerance mechanism of rice (Hussan et al., 2003). In some experiments, the damage of rice leaves is attributed to the accumulation of Na^+ in the shoot due to the transport of Na^+ from the root to the shoot under high external concentration (Lin et al., 2004).

3.2 Osmotic adjustments

Osmotic stress is one of the main limiting factors of crop production. The capability of osmotic adjustment is one of the basic characteristics of salinity tolerance of crops. The capability of osmotic adjustment at the cellular level dictates the capacity of the crop for salt tolerance (Bahaji et al., 2003). Salinity induces the accumulation of soluble sugars to reduce cell osmosis (Dubey and Singh, 1999). Some nitrogen compounds accumulate in response to salt stress and their accumulation in the rice roots is highly correlated with the Na^+ concentration (Nguyen et al., 2003). Mass proline accumulates in the rice leaves under salt stress (Zhang et al., 1997). The proline content of rice young seedlings under salinity stress is an important index of the degree of stress (Xie et al., 2005). The quantity of proline accumulation in the rice roots is a physiological index of the rice varieties for osmotic stress (Do et al., 2003).

Endogenous free ABA is accumulated rapidly in the seedlings of rice varieties after salt stress treatments. ABA accumulation levels are higher and maintained longer in the salt-tolerant varieties compared to those in sensitive varieties (Liu et al., 2003).

3.3 Transporters and trans-membrane movement

Molecular responses and signal transduction are induced under salinity stress. For instance, salinity stress induces gene expression of some important functional and regulatory proteins to protect plant cells from injury. In a wild rice species (*Oryza latifolia*), the thylakoid membrane shows high salt tolerance and it still maintains photosynthetic activity even under high NaCl conditions. The wild rice also has specific proteins, which may be partially responsible for salt tolerance (Nakamura et al., 2004).

The quantity of vacuolar Na^+/H^+ "antiporter" induced in rice plants under salt stress is one of the most important indexes in deciding salt tolerance capability (Fukuda et al., 2004). LEA proteins are accumulated during the salinity-triggered growth reduction of young seedlings whilst they are mobilized during the recovery of seedlings from salinity stress (Chourey et al., 2003). Na^+ and K^+ uptake in rice are mediated by different transporters such as *TaHKT1* for K^+

and Na^+ and *OsHKT1* for only Na^+. *OsHKT* transporters are involved in Na^+ movement in rice and *OsHKT1* specifically mediates Na^+ uptake in the rice roots when the plants are K^+ deficient (Garciadeblás et al., 2003).

4. GENETICS OF SALT TOLERANCE IN RICE

Genetic research shows that salinity tolerance in rice is a complicated phenomenon of various physiological reactions and a multigenic trait, which is controlled by many genes distributed on different chromosomes (Flowers, 2004). Inheritance variance rates of leaf death and seedling dry matter under salt stress show the additive and interactive effects of rice salt tolerance (Moeljopawira and Ikehashi, 1981). Salt tolerance at the rice seedling stage that is likely controlled by only a few genes shows mainly additive and dominant effects and has no epistatic interaction (Jones et al., 1985).

Akbar et al. (1985) studied the genetic variation of rice seedling characters under salt stress. The results showed that seedling height, Na^+ and Cl^- concentration of above-ground parts, and shoot, leaf and roots dry matter had strong additive effects and high heritability whilst Na^+ and Cl^- concentration of roots and root length had stronger dominant effects and lower heritability. Root Na^+ and Cl^- concentration and root length were controlled by at least three genes, whereas root, shoot and leaf length and shoot dry matter were controlled by two genes.

A BC_1 [Peta/Pokkali (salt-tolerant) × Peta (salt-sensitive)] population was employed to detect the QTLs for salt tolerance at different stages of rice growth. There are four putative QTLs for salt tolerance at the seedling stage and all of the positive alleles came from Pokkali. The favorite alleles of the QTLs for the salt tolerance at the maturity stage came from both parents. Two QTLs in the vicinity of RG678 and RZ400B-RZ792 exhibited salt tolerance over the whole growth period. The rice salt tolerance at seedling and maturity stages had common genetic bases (Gu et al., 2000). Salt-tolerant mutant lines have been obtained. Their salt-tolerant traits have been stably inherited for thirteen generations. The effect of a major gene may be present in these mutants.

In most of the above studies, the genetic analysis was made at the seedling stage. The responses of most indexes show that rice salt tolerance is a quantitative trait even though different rice varieties and indexes of salt tolerance have been used by the researchers. It may be concluded that the genetic variation of rice salt tolerance is mainly controlled by additive and dominant effects, which are the important genetic basis of salt tolerance at the seedling stage. With regard to the inheritance of salt tolerance in hybrid rice, some scholars believe that the traits for salt tolerance in the hybrid rice are not yet well known. The reason for this is still unclear.

5. MOLECULAR STUDIES OF SALT TOLERANCE IN RICE

5.1 Molecular mapping

The salinity tolerance in rice shows QTLs controlled by many genes (Akbar and Yabuno, 1977; Gong et al., 1999; Flowers, 2004). Lin et al. (2004) identified three QTLs on chromosomes 1, 6, and 7 for the survival rate of seedlings under salt stress. Among them, *SNC-7* and *SKC-1* were the main genes, which explained 48.5% and 40.1% of the observed phenotypic variances, respectively. Gu et al. (2000) found four QTLs associated with salt tolerance at the seedling stage and all of their alleles came from salt-tolerant varieties. Lin et al. (1998) detected one QTL for the survival period of seedlings on chromosome 5, accounting for 11.6% of the observed phenotypic variance. Seven QTLs, which were associated with the salt tolerance at the seedling stage, were detected on chromosome 5, 6, 7 and 10 (Prasad et al., 2000). Koyama et al. (2001) identified 11 QTLs, which were associated with Na^+ concentration, K^+ concentration, Na^+ uptake, K^+ uptake, and Na^+/K^+ ratios. Twenty-three QTLs for six traits were identified on the ten chromosomes, except chromosomes 5 and 10. These included five for SST, six for SDS, four for SKC, four for SNC, one for RKC, and three for RNC (Sun et al., 2007). Table 5.1 *summarizes* the findings of QTLs analysis for salt tolerance in rice.

Table 5.1 Status of QTL analysis for salt tolerance in rice

Salt tolerance trait	Mapping material	Population and marker	Number of QTLs (Chromosome)	Variance explained (%)	Reference
SDS	Nona Bokra×Koshihikari	$F_{2:3}$, RFLP	3(1, 6, 7)	13.9–18.0	Lin et al., 2004
NC	IR4630×IR15324	RIL (F_6), AFLP	2(4, 6)	6.4–19.6	Koyama et al., 2001
KC			2(1, 4)	8.8–10.6	
NU			1(1)	8.9	
KU			3(4, 6, 9)	6.8–19.6	
NKR			2(1, 4)	9.1–9.6	
DM			1(6)	9.7	
SG	IR64×Azucena	DH (F_1), RFLP	2(6, 7)	16.3–19.5	Prasad et al., 2000
SSL			1(6)	18.9	
SDM			3(5, 6, 10)	13.5–17.9	
SV			1(6)	16.3	

(continued)

Salt tolerance trait	Mapping material	Population and marker	Number of QTLs (Chromosome)	Variance explained (%)	Reference
RSI	Pokkali×Peta	BC_1, RFLP	2(5, 7)	—	Gu et al., 2000
NC			1(9)	—	
SFW/DMW			1(6)	—	
SW			3(1, 7, 10)	—	
SSR			1(3)	—	
MPL			1(5)	—	
PH			1(9)	—	
NSMP			1(6)	—	
SDS	Tesanai 2×CB	RIL (F_7), RFLP	1(5)	11.6	Lin et al., 1998
ASDS	ZYQ8×JX17	DH (F_1), RFLP	8 (1, 2, 2, 3, 7, 8, 8, 12)	10.2–38.4	Gong et al., 1999
ISSVF	M-20×77-170	$F_{2:3}$, RFLP	1(7)	—	Zhang et al., 1995
SST	IR64×Tarom Mola	BC_2 F_8, SSR	5(2, 3, 4, 7)	8.77–14.47	Sun et al., 2007
SDS			6(2, 3, 4, 7, 9)	8.87–13.03	
SKC			4(1, 2, 3, 8)	9.59–18.9	
SNC			4(2, 3, 11, 12)	9.91–11.5	
RKC			1(6)	9.55	
RNC			4(6, 7, 9)	9.89–14.75	

SDS: Survival days of seedling; NC:Na^+ concentration; KC: K^+ concentration; NU: Na^+ uptake; KU: K^+ uptake; NKR: Na^+/K^+ ratio; DM: Dry mass; SG: Seed germination; SSL: Seedling shoot length; SDM: Seedling dry matter; SV: Seedling vigour; RSI: Ranking of salt injury at the seedling stage; SFW/DMW: The ratio of shoot fresh weight/dry matter weight of the seedling; SW: Straw weight at mature stage; SSR: Seed setting rate; MPL: Main panicle length; PH: Plant height; NSMP: Number of spikelets in main panicle; ASDS: Average survival days of seedling; ISSVF: The individual of the sum of scores for vigour and fertility; SST: Score of salt toxicity of leaves; SKC: Shoot K^+ concentration; SNC: Shoot Na^+ concentration; RKC: Root K^+ concentration; RNC: Root Na^+ concentration.

5.2 Signal transduction

Salinity tolerance in plants is a complex process including the plant response to external stress, initiation of signal transduction, signal identification and transduction. The gene expression relative to salinity tolerance is adjusted by stress-induced signals. Hashimota et al. (2004) investigated the roots of two-week-old rice seedlings under salt stress considering a novel rice PR10 protein being specifically induced in the root by salt stress possibly via the jasmonic acid signalling pathway. The signal for salt stress induction was the expression of genes encoding antioxidant enzymes in the rice seedlings. This precise system of antioxidant adjustment keeps the environmental balance of oxidation-reduction in the cell to protect organelles from ion toxicity (Menezes-Benavente et al., 2004). Two rice transduction initiation factors OseIF5A-1 and OseIF5A-2 are induced and regulated by salt stress (Chou et al., 2004). Some research results show that the rice flavonoid pathway genes *OsDfr* and *OsAns* are induced by high salt stress responsive promoter elements that interact with the transcription activator Osc-MYB. The flavonoid signal transduction is possibly one of the important stress signal transduction pathways (Ithal and Reddy, 2004).

6. GENE CLONE AND TRANSFORMATION OF SALT TOLERANCE IN RICE

Great progress has been made in the identification of genetic resources and the isolation of useful genes in rice. Fifty-seven induced genes in rice were detected using cDNA microarrays and RNA imprinting technology (Rabbani et al., 2003). Some QTLs associated with salt tolerance traits have been located and cloned (Huang et al., 2003; Mohanty et al., 2002; Suprasanna, 2003; Hayashi et al., 1997; Qian et al., 2003). The aconitifolia-pyrroline-5-carboxylatelate synthetase gene was successfully transformed into *indica* rice cv IR-50, and the transgenic plants show high salt tolerance (Anoop and Gupta, 2003). Over-expression of a barley aquaporin gene in rice using transformation increases the shoot/root ratio and raises the salt sensitivity in the transgenic rice plants (Katsuhara et al., 2003). The *OPBP1* gene was transformed into japonica rice Xiushui 11 and shows improved resistance to salt stress. Under the same salt stress conditions, transgenic plants grow faster, have significantly higher biomass and chlorophyll content compared to non-transgenic plants (Li et al., 2006). Some important salt tolerance genes from wild rice (*Porteresia coarctata* T.) were transformed into cultivated rice using tissue culture and molecular biology approaches (Brar et al., 2002). The growth of shoot and root of the F_3 transgenic rice showed stronger tolerance under high salt stress.

7. BREEDING FOR SALINITY TOLERANCE IN RICE

The present section presents a succinct overview of the achievements made in addressing salinity tolerance in rice using *in vitro* techniques (doubled haploidy), genetic engineering and mutagenesis.

7.1 *In vitro* techniques for enhancing salinity tolerance in rice

Seventy-nine di-haploid lines were produced in rice through anther culture of the cross of two *indica* breeding lines. In 1996, some high-yielding salt-tolerant AC-derived lines (IR51500-AC11-1, IR51500-AC17, IR51485-AC6534-4, IR72132-AC6-1, IR69997-AC1, IR69997-AC2, IR69997-AC3 and IR69997-AC4) had been generated in just 3 years (Senadhira et al., 2002; Gregorio et al., 2002). The line IR51500-AC11-1 was released as a salt-tolerant cultivar in the Philippines with the name PSBRc50 or "Bicol" (Senadhira et al., 2002). This is the first F_1 AC-derived line from an *indica*/*indica* cross to be released as a cultivar for cultivation in saline-prone areas. In India, IR51500-AC17 and IR51485-AC6534-4 were named as commercial cultivars CSR21 and CSR28, respectively, for cultivation on salt-affected soils (IRRI, 1997). Pokkali and other salt-tolerant donors underwent cell culture to induce somaclonal variation. Somaclonal variants of Pokkali with improved agronomic traits were identified. The variant (TCCP 266-2-49-B-B-3) had desirable levels of all tested characteristics and retained salinity tolerance similar to Pokkali. It is superior to Pokkali as a donor of salt tolerance in hybridization programs (Gregorio et al., 2002). Gu et al. (2000) reported the development of semi dwarf mutants derived from the salt-tolerant landrace Pokkali via anther culture because some rice landraces exhibited greater tolerance to salt but they were agronomically unacceptable because of their tall stature.

7.2 Genetic transformation for salinity tolerance in rice

Transgenic rice expressing the *codA* gene had enhanced salt tolerance and could recover to normal growth at a faster rate than the wild type after an initial growth inhibition under salt stress. Overproduction of proline in transgenic rice increased biomass production and enhanced flower development under salinity stress. Hoshida et al. (2000) examined the potential role of the photorespiration in the protection against salt stress with transgenic rice. Roy and Wu (2001) reported that the introduction of oat arginine decarboxylase (ADC) into rice led to salt stress-induced up-regulation of ADC activity and polyamine accumulation in the transgenic rice plants. All transgenic rice plants showed increased biomass under salinity stress conditions. Flowers (2004) also reviewed the work of Garg et al. (2002) involving the transformation of rice to overexpress genes that led to the synthesis of trehalose. SNAC1-overexpressing transgenic rice showed enhanced salt tolerance (Hu et al., 2006).

7.3 Mutation induction in rice for salinity tolerance

Six mutants with higher salt tolerance than the parent variety Bicol and two salt-tolerant varieties derived from the salt-sensitive variety IR29 were obtained in the year 2001. The development of useful salt-tolerant mutants from IR29 with superior plant vigour is a major contribution to addressing the scourge of salinity in the agricultural lands of Southeast Asia because this variety is well-adapted and widely cultivated by farmers in this region. One mutant variety Shua-92 and two rice mutants derived through mutation breeding from the two standard varieties IR8 and Pokkali, were evaluated over two years for their yield performance in salt-affected soils with pH from 7.63 to 7.68 and EC_e from 7.11 dS m^{-1} to 8.0 dS m^{-1}. The mutant variety Shua-92 produced 40% and 49% higher paddy yield on salt-affected soils than the well-known salt-tolerant varieties Nona Bokra and Pokkali, respectively (Baloch et al., 2003).

8. FUTURE PROSPECTS

8.1 Study on exploitation and utilization of rice germplasm for salt tolerance

Rice germplasm resources are the available gene pool and the basic material utilized for rice breeding and cultivar improvement. The capability of salt tolerance in rice is mainly dictated by its genetic characteristics. The narrow genetic background of the salt-tolerant parents is an important limiting factor in rice breeding for salt tolerance, which is not yet resolved. It is necessary to exploit rice materials with high salt tolerance taken from the present rice germplasm for further breeding work.

8.2 Research on physiological mechanisms for salt tolerance in rice

Both osmoregulation and ion uptake/alteration affect the capability of salt tolerance in rice. The pathways for this regulation are still poorly understood. There are several opinions about the roles of the Na^+ and proline in the salt stress processes (Flowers and Yeo, 1981; Zhang et al., 1997). The functions of the trans-membrane channel proteins in the process of stress response and the ion transporters in the membranes are unknown. In addition the responses to salt tolerance in rice are different between the tissue and the whole plant level. Therefore, these issues need to be comprehensively researched so that the physiological mechanisms for salt tolerance in rice are better understood.

8.3 Study of rice salt tolerance using molecular markers mapping

Some QTLs associated with the rice salt tolerance have been identified (Gong et al., 1999; Prasad et al., 2000; Lin et al., 2004). As scientists use different populations, experimental mapping approaches and statistical analysis methods, the number of QTLs and their location in the chromosomes are different. Consequently, some selected QTLs, well identified, with stable inheritance and with consistently observed phenotypic variance, should be utilized to analyze the salt tolerance over several years and places. These QTLs can be further used in rice breeding for saline-alkaline tolerance by means of marker-aided selection.

8.4 Study of signal transduction for salt tolerance in rice

Under salt stress, there is a need to study the whole process and multi-pathways of molecular response and signal transduction, especially the factors of gene expression and promoters, and to build a genetic network to unravel the basic elements of signal transduction of salt tolerance in rice.

8.5 Study on core germplasm for salt tolerance in rice

Conventional breeding methods and current molecular biology approaches should be used to characterize and evaluate thoroughly rice elite germplasm to decipher salt tolerance genetics. This work involves approaches of conventional crossing, tissue culture, interspecific hybridization, marker-aided selection, transgene technology and so on. Strongly salt-tolerant rice germplasm will then be utilized quickly in rice breeding for improved rice production in salt-affected lands.

REFERENCES

Akbar, M., & Yabuno, T. (1977). Breeding for saline resistant varieties of rice, IV. Inheritance of delayed-type panicle sterility induced by salinity. *Jpn. J. Breed*, *27*, 237-240.

Akbar, M., et al. (1985). Genetics of salt tolerance in rice, Rice Genetics, Proceedings International Rice Genetics Symposium. *IRRI*, *5*, 399-409.

Anoop, N., & Gupta, A. K. (2003). Transgenic *indica* rice cv IR-50 over-expressing *Vigna aconitifolia* Δ^1-pyrroline-5-carboxylatelate synthetase cDNA shows tolerance to high salt. *J. Plant Biochem. Biotechnol*, *12*(2), 109-116.

Baba, T., & Fujiya, M. H. (2003). Short-term response of rice and tomato to NaCl stress in relation to ion transport. *Soil Sci. Plant Nutr*, *49*(4), 513-519.

Bahaji, A., et al. (2003). Uptake of an endocytic marker by rice cells: variations related to osmotic and saline stress. *Plant Cell Physiol*, *44*(10), 1100-1111.

Baloch, A. W., et al. (2003). Induction of salt tolerance in rice through mutation breeding. *Asian J. Plant Sci*, *2*, 273-276.

Brar, D. S., et al. (2002). *Transferring agronomically important genes from wild species into rice: application of tissue culture and molecular approaches*, Abstracts International Conference on Wild Rice, Katmandu, Nepal, 17-18.

Brugnoli, E., & Lauteri, M. (1991). Effects of salinity on stomatal conductance, photosynthetic capacity, and carbon isotope discrimination of salt-tolerant and salt-sensitive C_3 non-halophytes. *Plant Physiol*, *95*, 628-635.

Chourey, K., et al. (2003). Accumulation of *LEA* proteins in salt (NaCl) stressed young seedling of rice (*Oryza sativa* L.) cultivar Bura Rata and their degradation during recovery from salinity stress. *J. Plant Physiol*, *160*(100), 1165-1174.

Chou, W. C., et al. (2004). Expression of genes encoding the rice transaction initiation factor EIF5A is involved in development and environmental responses. *Physiol. Plant*, *121*(1), 50-57.

Do, T. H., et al. (3003). Proline accumulation and Δ-pyrroline-5-carboxylate synthetase gene properties in three rice cultivars differing in salinity and drought tolerance. *Plant Sci*, *165*(5), 1059-1068.

Dubey, R. S., & Singh A. K., (1999). Salinity induces accumulation of soluble sugars and alters the activity of sugar metabolising enzymes in rice plants. *Biol Plant*, *42*(2), 233-239.

Flowers, T. J. (2004). Improving crop salt tolerance. *J. Exp. Bot*, *55* (396), 307-319.

Flowers, T. J., & Yeo, A. R. (1981). Variability in the resistance of sodium chloride salinity within rice varieties. *New Phytol*, *88*, 363-373.

Fukuda, A., et al. (2004). Function, intracellular localization and the importance in salt tolerance of a vacuolar Na^+/H^+ antiporter from rice. *Plant Cell Physiol*, *45*(2), 146-159.

Garciadeblás, B., et al. (2003). Sodium transport and *HKT* transporters: the rice model. *Plant J*, *34*(6), 788-801.

Garg, A., et al. (2002). Trehalose accumulation in rice plants confers high tolerance levels to different abiotic stresses. *Proc. Natl. Acad. Sci. USA*, *99*, 15898-15903.

Cheng, G. Y., et al. (1994). Study on Na_2CO_3 screening concentration and determinative index during rice seedling stage. *J. of Yanbian Agric. Coll*, *6*(1), 42-46.

Cheng, G. Y., et al. (1995). Study on the saline-Alkaline tolerance of rice varieties II. A screening method for Alkaline-tolerance of rice varieties. *J. of Yanbian Agric. Coll*, *17(*4), 195-201.

Cheng, G. Y., et al. (1996). Study on the Saline-Alkaline Tolerance in Plants (1) Correlation Analysis between Saline-tolerance and Alkaline-tolerance of Rice. *J. of Jilin Forestry Univ*, *12*(4), 30-33.

Gong, J. M., et al. (1999). Comparative study on QTLs controlling important agronomic traits of rice under salt stress and nonstress Environment. *Sci in China Series C: Life Sciences*, *21*, 46.

Greenway, H., & Munns, R. (1980). Mechanisms of salt tolerance in non-halophytes. *Annu. Rev. Plant Physiol*, *31*, 149-190.

Gregorio, G. B., et al. (2002). Progress in breeding for salinity tolerance and associated abiotic stresses in rice. *Field Crops Res*, *76*, 91-101.

Cssri, I., & Girdhar, I. K. (1988). Effect of saline irrigation water on the growth, yield and chemical composition of rice crop gorwn in saline soil. *J. Indian Soc. Soil Sci*, *26*, 324-329.

Gu, X. Y., et al. (2000). Preliminary detection of quantitative trait loci for tolerance in rice. *Chin. J. Rice Sci*, *14*(2), 65-70.

Guo, W. M., et al. (2004) Study on Selection of Salt Tolerance Indexes in Rice Bud and Seedling Stage. *Zhejiang Agric. Sci*, *1*, 30-33.

Guo, Y., et al. (1997). Salt tolerance rice mutant lines controlled by a major effect gene were obtained by cell engineering technique. *Acta Genet. Sin*, *24* (2), 1221-126.

Han, Z. H., et al. (1998). Effect of NaCl on the imbibed seed germination and seedling growth of different cultivars of rice. Zhiwu Shenglixue Tongxun, *34*(5). 339-342.

Hashimota, M., et al. (2004). A novel rice PR10 Protein, *RsOsPR10*, specifically induced in roots by biotic and abiotic stresses, possibly via the jasmonic acid signaling pathway. *Plant Cell Physiol*, *45*(5), 501-559.

Hayashi, H., et al. (1997). Transformation of *Arabidopsis thaliana* with the *codA* gene for choline oxidase; accumulation of glycine betaine and enhanced tolerance to salt and cold stress. *Plant J*, *12*, 133-142.

Hu, H., et al. (2006). Overexpressing a NAM, ATAF, and CUC, (NAC). transcription factor enhances drought resistance and salt tolerance in rice. *Proc. Natl. Acad. Sci. USA*, *103*(35), 12987-12992.

Huang, J., et al. (2003). Molecular cloning and characterization of rice 6-phosphogluconate dehydrogenase gene that is up-regulated by salt stress. *Mol. Biol. Rep*, *30*(4), 223-227.

Hussan, N., et al. (2003). Mechanism of salt tolerance in rice, *Pedosphere*, *13*(3), 231-238.

IRRI. (1997). *Program report for 1996*, International Rice Research Institute, Philippines.

Ithal, N., & Reddy, A. R. (2004). Rice flavonoid pathway genes *OsDfr* and *OsAns*, are induced by dehydration, high salt and ABA and contain stress responsive promoter elements that interact with the transcription activator. *Osc-MYB*, *Plant Sci*, *166*(6), 1505-1513.

Jones, M. P., et al. (1985). Genetic analysis of salt tolerance in mangrove swamp rice, Rice Genetics, Proceeding of International Rice Genetics Symposium. *IRRI*, *5*, 411-422.

Katsuhara, M., et al. (2003). Over-expression of a barley aquaporin increased the shoot/root ratio and raised salt sensitivity in transgenic rice plants. *Plant Cell Physiol*, *44*(12), 1378-1383.

Khan, M. A., & Abdullah, Z. (2003). Salinity-sodicity induced changes in reproductive physiology of rice under dense soil conditions. *Environ. Exp. Bot*, *49*(2), 145-157.

Khan, M. S. A., et al. (1997). Effect of sodium chloride on germination and seedling characters of different types of rice. *J. Agron. Crop Sci*, *179*, 163-169.

Koyama, M. L., et al. (2001). Quantitative trait loci for component physiological traits determining salt tolerance in rice. *Plant Physiol*, *125*, 406-422.

Lee, C. K., et al. (2002). Growth and yield of rice as affected by saline water treatment at different growth stages. *Korean J Crop Sci*, *47*(6), 402-408.

Li, H. B., et al. (2006). Responses of rice leaf photosynthetic parameters to light intensity under NaCl stress. *Chin. J. Appl. Ecol*, *17*(9), 1588-1592.

Li, N. Y. & Guo, Z. J. (2006). Overexpression of two different transcription factors, OPB P1 and OsiWRKY enhances resistance against pathogen attack and salt stress in rice. *Chin. J. Rice Sci*, *20*(1), 13-18.

Liang, Z. W., et al. (2004). Effect of the main growth characteristics of rice under saline-alkali stress. *Ecol. and Environ*, *13*(1), 43-46.

Lin, H. X., et al. (1998). Identification of QTL for salt tolerance in rice via molecular markers. *Chin. J. Rice Sci*, *12*(2), 72-78.

Lin, H. X., et al. (2004). QTLs for Na^+ and K^+ uptake of the shoots and roots controlling rice salt tolerance. *Theor. Appl. Genet*, *108*, 253-260.

Liu, C. L., et al. (2003). Multiple tolerance of rice to abiotic stresses and its relationship with ABA accumulation. *Acta Agron. Sin*, *29*(5), 725-729.

Menezes-Benavente, L., et al. (2004). Salt stress induces altered expression of genes encoding antioxidant enzymes in seedlings of a Brazilian *indica* rice (*Oryza sativa* L.). *Plant Sci*, *166*(2), 323-331.

Moeljopawira, S., & Ikehashi, H. (1981). Inheritance of salt tolerance in rice. *Euphytica*, *30*, 230-291.

Mohanty, A., et al. (2002). Transgenics of an elite *indica* rice variety Pusa Basmati 1 harbouring the *codA* gene are highly tolerant to salt stress. *Theor. Appl. Genet*, *106*(1), 51- 57.

Munns, R. (2002). Comparative physiology of salt and water stress. *Plant*, *Cell Environ*, *25*, 239-250.

Nakamura, I., et al. (2004). Salt tolerance of the chloroplast thylakoid membrane in wild Oryza species, *Oryza latifolia* Desv. *Jpn. J. Crop Sci*, *73*(1), 84-92.

Nguyen, T. T. H., et al. (2003). Accumulation of some nitrogen compounds in response to salt stress and their relationships with salt tolerance in rice (*Oryza sativa* L) seedling. *Plant Growth Regul*, *41*(2), 159-164.

Prasad, S. R., et al. (2000). Molecular mapping of quantitative trait loci associated with seedling tolerance to salt stress in rice (*Oryza sativa* L). *Curr. Sci*, *78*, 162-164.

Qian, Q., et al. (2003). Isolation expression characteristics and chromosomal locations of three cDNA fragments under salt stress in rice. *Acta Bot. Sin*, *45*(9), 1090-1095.

Rabbani, M. A., et al. (2003). Monitoring expression profiles of rice genes under cold drought and high-salinity stresses and abscisic acid application using cDNA microarray and RNA gelblot analyses. *Plant Physiol*, *133*(4), 1755-1767.

Roy, M., & Wu, R. (2001). Arginine decarboxylase transgene expression and analysis of environmental stress tolerance in transgenic rice. *Plant Sci*, *160*(5), 869-875.

Senadhira, D., et al. (2002). Development of the first salt-tolerant rice cultivar through *indica/indica* anther culture. *Field Crop Res*, *76*, 103-110.

Shereen, A., et al. (2002). Rice cultivation in saline soil (Chapter 20), In Ahmad, R., & Malik, K. A. (Eds.), *Prospects for saline agriculture*, Kluwer Academic Publishers, Dordrecht, the Netherlands, 189-192.

Shu, L.B., et al. (2007). Advance in Rice Proteomics of Drought-and Salt Tolerance. *Biotechnology Bulletin*, *4*, 31-37.

Su, J., & Wu, R. (2004). Stress-inducible synthesis of proline in transgenic rice confers faster growth stress conditions than that with constitutive synthesis. *Plant Sci*, *66*(4), 941-948.

Suprasanna, P. (2003). Building stress tolerance through over-producing trehalose in transgenic plants. *Trends Plant Sci*, *8*(8), 355-357.

Sun, Y., et al. (2007). Mining favorable salt-tolerant QTL from rice germplasm using a backcrossing introgression line population. *Acta Agron. Sin*, *33*(10), 1611-1617.

Wang, Z. S., et al. (1986). Physiological studies on salt tolerance in rice I. varietal differences in plant water relation and osmoregulation under salt stress. *J. of Agric. Sci*, *2*(3), 1-11.

Wang, R. L., et al. (2002). Effect of salt stress on photosynthetic characteristics in rice. *J. of Nanjing Agric. Univ*, *25*(4), 11-14.

Xie, G. S., et al. (2005). Comparison of the response of rice seedlings to NaCl and $NaHCO_3$ stress with different pH value. *J. of Huazhong Agric.Univ*, *24*(2), 121-124.

Yan, B., et al. (1995). The study on increasing salt resistance of rice by calcium. *Acta Agron. Sin*, *21*(6), 685-690.

Zhang, G. Y., et al. (1995). RFLP tagging of a salt tolerance gene in rice. *Plant Sci*, *110*, 227-234.

Zhang, H., et al. (1997). A physiological study on the salt-tolerant mutant of rice. *Acta Phytophysiol. Sin*, *23*(2), 181-186.

Zhang, R. Z., et al. (2006). Effect of saline-alkali stress on source-sink and yield of rice. *Chin. J. Rice Sci*, *20*(1), 116-118.

Zhao, K. F. (2002). The plant adaptation to salt stress. *Biol. Bull*, *37*(6), 7-10.

Zheng, S. H., et al. (1996). Distribution of Na^+ and Cl^- in the roots of different rice genotypes under salt stress. *J. South China Agric. Univ*, *17*(4), 24-28.

CHAPTER 6.

ADVANCES IN BREEDING BARLEY FOR SALT TOLERANCE

Chengdao Li[1], Meixue Zhou[2], Zhonghua Chen[3] and Guoping Zhang[4]

[1]Centre for Crop and Food Innovation, Food Futures Institute, Murdoch University, 90 South Street, Murdoch Western Australia 6150;

[2]Tasmanian Institute of Agricultural Research, University of Tasmania, P.O.Box 46, Kings Meadows, Tas 7250, Australia;

[3]Institute of Biomedical and Life Sciences, Bower Building, University of Glasgow, Glasgow, G12 8QQ, UK;

[4]College of Agriculture and Biotechnology, Zhejiang University, Huajiachi Campus, Hangzhou 310029, China

1. INTRODUCTION

Almost three-quarters of the surface of the world is covered by the ocean's salt water and so it is not surprising that salt is dramatically dominant on the earth's land surface. Moreover, soil salinization in the arable land is ever-increasing due to the expansion of irrigated agriculture into arid zones with high evapotranspiration rates as well as poor irrigation and drainage practices (Flowers et al., 1977). Salinity is one of the main problems that negatively affect soil fertility and physical properties, thus severely affecting the growth and economic yield of many important crops (Maas and Hoffman, 1977). Salinity can be alleviated through either soil reclamation or growing salt-tolerant crops. However, soil reclamation is a very expensive process, and hence the cultivation of salt-tolerant crop species and varieties is the most practical solution when the salinity level is relatively low. It is well known that there are significant genotypic differences with respect to salt tolerance between and within plant species (Rana, 1986; Suhayda et al., 1992; Zhong et al., 1995). Among the main cereal crops, barley is highly tolerant to salinity, thus offering a means for the efficient utilization of saline soils and improvement of agricultural productivity in these environments. However, barley is also affected by salt toxicity in many areas of the world. As mentioned above, distinct genetic differences in summarizes can be found among and within the barley species, thus providing the fundamental basis for improvement of summarizes. This chapter summarizes progress on physiology, screening methods, molecular genetics, and genetic improvement of barley for salinity tolerance through conventional and biotechnological approaches.

2. SALINITY TOXICITY AND PHYSIOLOGICAL MECHANISMS FOR SALT TOLERANCE IN BARLEY

2.1 Salinity toxicity

The presence of high concentrations of Na^+, Cl^-, Mg^{2+}, and SO_4^{2-} in saline soils inhibits growth of many plants. The main problems associated with high salinity are 3-fold: 1) A high salinity is associated with a low soil water potential, giving rise to symptoms similar to those of water stress (osmotic effects); 2) Concentrations of specific ions, especially Na^+, Cl^-, may be toxic to plants; and 3) High levels of NaCl may give rise to an ion imbalance (mainly calcium and potassium), and lead to deficiency symptoms.

Toxicity effects of salinity may also include inhibition of nitrate uptake by chloride, probably because both ions are transported across the plasma membrane by the same carrier. High Na^+ may replace Ca^{2+} on the root cell membranes, which may give rise to leakage of H^+ from the root cells. It may also reduce the influx and enhance the efflux of calcium. The decreased influx probably results from competition for binding sites in the cell wall, which decreases the concentration in the protein of the plasma membrane responsible for calcium influx. The toxicity of specific ions may subsequently lead to an ion imbalance and nutrient deficiency, especially calcium deficiency (Rengel, 1992).

Ion imbalance will occur when the plants are exposed to salinity stress, thus resulting in nutritional disorders, i.e. excessive accumulation of some nutrient elements or deficiency of some other elements. In most cases, salinity stress reduces the uptake and accumulation of K^+, leading to an imbalance of K^+/Na^+ ratio, which has been considered as the major reason of salinity toxicity to plants. Leonova et al. (2005) found that soil salinization increased the sodium content in barley seedlings as compared to the control plants. In general, salt-sensitive cultivars accumulated more Na^+ in their shoots than salt-tolerant cultivars whilst the reciprocal pattern was found in the roots. On the other hand, soil salinization decreased K^+ content in the shoots of the salt-sensitive cultivar as compared to the control whereas in the most tolerant cultivars, the potassium content increased. Gorham et al. (1990) comparing the differences of K^+ and Na^+ concentrations in the plant tissues of the *Triticeae* species found out: i) *Triticum aestivum* L., *Secale cereale*, and *Aegilops squarrosa* had the lowest leaf Na and highest leaf K concentrations; ii) *Triticum durum* and the *Hordeum* species did not have this character; and iii) the better growth of *Hordeum vulgare* than that of *Triticum durum* under similar salt concentrations in the youngest fully-expanded leaves may be attributed to a better compartmentation of Na^+, Cl^-, and K^+ among the plant tissues or between different compartments within the cells.

2.2 Physiological mechanisms for salinity tolerance

Salinity tolerance is defined as the ability of the plant to survive under salinity and complete the growth cycle with an acceptable growth and yield. It is a complex trait involving different mechanisms. Plants employ various strategies in response to salinity as follows: (i) minimizing the initial Na^+ entry and Na^+ loading to the xylem; (ii) maximizing the Na^+ efflux from the root, its recirculation out of the shoot, intracellular compartmentation or allocation to old leaves; (iii) maintaining a relatively high cytosolic K^+/Na^+ ratio; (iv) accumulating/synthesising optimal amounts of compatible solutes; and (v) increasing enzymatic and non-enzymatic antioxidant defence systems (Bohnert et al., 1995; Hasegawa et al., 2000; Tester and Davenport, 2003).

For most plants, Na^+ is not an essential nutrient, although low Na^+ concentrations often stimulate plant growth in many species. To a large extent this is attributed to the role of Na^+ as an osmoticum in the vacuole, reducing the need for K^+ (Marschner, 1995). Sodium has been shown to be essential for maximal growth of certain halophytic C_4 plant species, such as bladder saltbush (*Atriplex vesicaria*), fire bush (*Kochia childsii*), proso millet (*Panicum miliaceum* L.), and saltgrass (*Distichlis spicata* L.) (Brownell and Crossland, 1972; Flowers et al., 1977; Subbarao et al., 2003). In C_4 species, Na^+ is considered to be a beneficial element and, to some extent, can replace certain K^+ functions such as an internal osmoticum, stomatal function, photosynthesis, counter-ion in long-distance transport, and enzyme activation (Marschner, 1995; Cramer, 1997; Subbarao et al., 2003). However, in glycophytes, elevated Na^+ levels are detrimental, causing specific ion toxicity and negatively affecting root nutrient uptake, especially ions such as K^+ and Ca^{2+} (Verslues et al., 2006).

Land plants do not appear to have specific transport systems for Na^+. Under high external Na^+ concentration, Na^+ enters cells passively via several routes (Cheeseman, 1982; Xiong and Zhu, 2002). This initially high, unidirectional influx of Na^+ into roots is found in species like barley (Kronzucker et al., 2006), wheat (Davenport, 1998), maize (Jacoby and Hanson, 1985;

Zidan et al., 1991) and *Arabidopsis* (Elphick et al., 2001; Essah et al., 2003). This influx is mostly mediated by non-selective cation channels (NSCCs) (Demidchik et al., 2002), whose usual function is related to the uptake of other cations such as K^+ (Uozumi et al., 2000) and Ca^{2+} (White and Davenport, 2002). The unidirectional Na^+ influx rates can exceed net Na^+ uptake rates by an order of magnitude, implying the involvement of high Na^+ efflux (Davenport et al., 1997; Essah et al., 2003).

Na^+ competes with K^+ for binding sites, which are essential for the cellular function. With over 50 enzymes activated by K^+, this phenomenon disrupts numerous enzymatic processes in the cytoplasm (Bhandal and Malik, 1988; Marschner, 1995). For example, the protein synthesis requires high concentrations of K^+ for the binding of *t*RNA to the ribosomes, but this is inhibited by high Na^+ *in vitro* (Hall and Flowers, 1973; Jones et al., 1979). Enzymes including malate dehydrogenase, aspartate transaminase, glucose 6-P dehydrogenase, and isocitrate dehydrogenase isolated from the salt-sensitive broad bean (*Phaseolus vulgaris* L.) and the salt-tolerant pop saltbush (*Atriplex spongiosa* F. Muell.) and glasswort (*Salicornia australis* Sol. ex F. Muell.) are equally sensitive to Na^+ up to 500 mmol L^{-1} *in vitro* (Greenway and Osmond, 1972). Thus, it is essential for plants to employ all kinds of strategies to reduce salt toxicity. Among these, Na^+ exclusion and vacuolar compartmentation are crucial for plant salt adaptation.

Exclusion of Na^+ from the cytosol has been suggested to be a key mechanism for salt tolerance in plants (Schubert and Läuchli, 1990; Tester and Davenport, 2003). Physiological mechanisms of exclusion that operate at the cellular and whole-plant level have been extensively reviewed (Greenway and Munns, 1980; Schachtman and Liu, 1999; Munns, 2002; Véry and Sentenac, 2002; Shabala, 2003). Ion exclusion mechanisms could provide a degree of tolerance to relatively low concentrations of NaCl but not at high salinity (Yamaguchi and Blumwald, 2005). Salt-tolerant wild *Hordeum* species had better Na^+ excluding ability than cultivated barley (Garthwaite et al., 2005). When grown in 50 mmol L^{-1} NaCl, bread wheat excluded around 98% Na^+, but barley, durum wheat, and rice excluded about 94% (Munns, 2005). There is a strong correlation between salt exclusion and salt tolerance in cereals such as barley, rice and wheat (Flowers and Yeo, 1986; Chhipa and Lal, 1995; Ashraf and Khanum, 1997; Munns and James, 2003). In contrast to the Na^+ excluding ability of roots of rice, barley has the ability to prevent root to shoot Na^+ translocation at high external NaCl (Nakamura et al., 1996). However, a highly salt-tolerant wild relative of tomato (*Lycopersicon esculentum* Mill.) accumulates higher concentrations of Na^+ than the salt-sensitive domesticated tomato (Santa-Cruz et al., 1999). The sensitivity of a wild-type *Arabidopsis* and some mutants does not appear to be closely related to the shoot levels of Na^+ (Zhu et al., 1998; Nublat et al., 2001; Essah et al., 2003). Therefore, the correlation between plant salt tolerance and its ability to exclude Na^+ from uptake is not straightforward as initially believed.

Compartmentation of Na^+ into the vacuole is also vital for the growth and survival of halophytes and of many non-halophytes, such as barley. Barley is more tolerant to salt than wheat partially due to its greater ability to sequester Na^+ in the vacuole (James et al., 2006). This mechanism would avoid toxic effects of salt on photosynthesis and other key cytosolic metabolic processes (Maathuis et al., 1992; Fricke et al., 1996; Blumwald and Gelli, 1997; James et al., 2006). However, such Na^+ compartmentation is not sufficient unless the plant also possesses the ability to efficiently retain K^+ in the cytosol.

Excessive accumulation of salt ions are toxic to plant cells, thus they must be separated from the metabolic machinery of the cells if the toxic effects should be alleviated or avoided. In general, this is achieved by ion compartmentation, storing salt ions in vacuoles, where there is very weak metabolic activity (Flowers and Yeo, 1986). A possible survival strategy of plants under saline conditions is to sequester absorbed Na^+ in the vacuole, thus maintaining a higher K^+/Na^+ ratio in the cytoplasm (Greenway and Munns, 1980). The K^+/Na^+ antiport in the vacuolar membranes transports Na^+ from the cytoplasm to the vacuoles using a pH gradient generated by H^+-ATPase and H^+-PPase, which is considered to be related to the salt tolerance of plants (Atsunori et al., 1998). It has been demonstrated that the proton pump and the K^+/Na^+ antiport in the vacuolar membranes were important in ion selective absorption and compartmentation of Na^+ in barley seedlings (Garbarino and Dupont, 1988). Takuji et al. (1996) investigated mechanisms of salt tolerance in terms of the ATPase activity in roots, the Na excluding ability of roots, and the regulation of Na translocation from roots to the shoot by comparing barley and rice grown at high NaCl concentrations. It was found that the high salt tolerance of barley was ascribed to the inhibition of Na translocation from roots to shoot, and less reduction of ATPase activity in the roots. ATPase activities in the plasma membrane and tonoplast of barley roots were higher than those of rice when roots were exposed to NaCl stress. Recently, Ershov et al. (2005) confirmed that the changes in the activity of ion transporters under salt stress conditions correlated with the barley cultivar-specific tolerance to elevated NaCl concentrations.

In order to maintain water uptake in saline environments, plants will reduce the water potential, which is realized by accumulating solutes and synthesizing some new chemicals. Within the cytoplasm the osmotic adjustment is influenced by compatible solutes such as glycinebetaine, mannitol and proline. When a plant is exposed to salt stress, it responds initially to the changed water conditions by lowering the external water potential. These initial effects of salinity are probably the same for the cultivars differing in salt tolerance. However, when ion accumulation lasts for some time, differences in salt tolerance among genotypes appear (Munns, 1993), with the sensitive cultivars accumulating ions more quickly than the tolerant ones.

Ions enter plant cells through a trans-membrane protein. The process is driven by energy-consuming ion pumps, which use the energy stored in ATP to move protons by generating a difference of hydrogen ion concentration (pH) and electric potential (ΔE). It is assumed that Na^+ is 'mistaken' for potassium by a K^+ carrier or channel, but it is also possible that Na^+ enter cells through non-selective cation channels (Maser et al., 2002). It has been reported that a specific phospholipid environment is required for optimal ATPase activity, and changes in the phospholipids and free sterols of the cell membranes may contribute to salt tolerance (Norberg and Liljenberg, 1991; Mansour et al., 1994). Yamaguchi and Kasamo (2001) found that exogenously added tonoplast phospholipids would stimulate the activity of purified tonoplast H^+-ATPase. Meanwhile, fatty acids are considered to be important in salt tolerance of plants and micro-organisms (Somerville, 1995; Malkit et al., 2002). By using genetic mutants, Allakhverdiev et al. (1999) have demonstrated that unsaturated fatty acids in the membrane lipids could protect the photosynthetic machinery against salt stress-induced damage. Zhao and Qin (2005) found that addition of linoleic acid (LA) at 1 mmol L^{-1} into the solution culture possessed protective effects on the root tonoplast function against salt stress in barley seedlings; and this effect was accompanied with a significant suppression of the phospholipids and polyamines (PAs) degradation in the tonoplast vesicles. Moreover, these ameliorating effects of linoleic acid on the tonoplast function were also indicated by the increase in H^+-ATPase and H^+-PPase activities. An application of LA under saline conditions increased the activity of a vacuolar K^+/Na^+ antiport.

Polyamines (PAs) have been found to be related to salt tolerance in plants. The poly-cationic nature of PAs at physiological pH enhances their ability of mediating biological activity. They are able to bind negatively charged molecules, such as DNA (Basu et al., 1990), membrane phospholipids and proteins (Tassoni et al., 1996), and pectic polysaccharides (D' Oraci and Bagni, 1987). In addition, PAs can also be covalently bound to some specific proteins catalysed by a class of enzymes known as trans-glutaminases (TGase) to form bound PAs (Serafini-Fracassini et al., 1995). Both free and bound PAs associated to the tonoplast vesicles from barley seedlings were detected, and their contents were found to be closely related to salt tolerance of the plants (Zhao et al., 2000).

The normal calcium metabolism in plants is quite important to develop salt tolerance. Lynch and Läuchli (1985) investigated the influence of salt stress on the calcium nutrition in barley. Soil salinity decreased the Ca contents of shoots of two cultivars of barley grown under field conditions, particularly in younger leaves. Reduction in Ca content was more pronounced in the salt-sensitive cultivar than in the salt-tolerant cultivar. In the seedlings grown in solution culture, relatively low concentrations of NaCl inhibited the transport of Ca to the shoot and the inhibition was not due to effects of NaCl on Ca influx into the root or on transpiration. It was found that

NaCl inhibited Ca transport from root to shoot by interfering with the release of Ca into the root xylem, possibly via an effect on the active loading of Ca into xylem vessels. Similarly, Cramer et al. (1989) reported that salt-stressed plants often show symptoms of Ca deficiency. Calcium transport and tissue concentrations were markedly inhibited by salinity. The transport and tissue concentrations of Na were significantly affected by supplemental Ca. There were significant Na-Ca interactions with ion transport, ion accumulation, and growth. Lynch et al. (1988) proposed that the leaf growth in salt-stressed barley plants was reduced by sub-optimal Ca availability in the leaf meristem. One possible cause of such reduced Ca availability is that Na replaces Ca in the leaf apoplast (Zid and Grignon, 1985). Salinity stress has been shown to stimulate a release of Ca from intracellular compartments (Lynch and Lauchli, 1988). Calcium transport to the shoot is reduced in NaCl-stressed plants (Lynch and Lauchli, 1985; Wolf et al., 1990), and indeed, the ability to transport Ca to the shoot during salt stress has been proposed as an index of salt tolerance (Lahaye and Epstein, 1971). According to Tobe et al. (2003), the Ca present in saline soils would alleviate the toxic effects of other salt components on seed germination of barley.

In addition, ultra-structural alterations occur in the root cells of some species including barley when the plants are exposed to salt stress. Kramer (1984) suggested that the appearance of various alterations under salt stress might have a function in the adaptation of plants to salinity. Huang and van Steveninck (1990) observed that structural changes occurred in the meristematic cells of barley in response to moderate salinity stress. In the apical region of the root, salinity stress caused an increase in vacuolation, which may provide a means for the accumulation of excess ions. Salt treatment also induced the formation of many plastids in the cortical cells, often appearing to enclose part of the cytoplasm, which was then less dense than the surrounding cytoplasm. It was suggested that the plastid morphology might allow or alternatively result from adaptive changes in the protein synthesis or cytoplasmic composition.

3. PHENOTYPIC SCREENING METHODS FOR SALINITY TOLERANCE

An efficient breeding program for salinity tolerance requires new genetic resources but breeding efforts are constrained by a shortage of adequate field and laboratory screening tests (Zhu, 2000). Unfortunately, few screening procedures have proven successful for identifying salinity tolerance due to a number of reasons (Shannon, 1997). Firstly, screening a large number of genotypes for salinity tolerance is difficult due to the complexity and polygenic nature of salinity tolerance, which involves responses to cellular osmotic, ionic and oxidative stresses (Shannon, 1997; Zhu, 2000). Secondly, evaluating field performance by measuring yield under saline conditions is difficult because of the variability of salinity within fields (Richards, 1983; Daniells et al., 2001) and the interactions of salinity with other environmental factors (Shannon and Noble, 1990; Flowers, 2004). Thus, physiological traits measured under controlled conditions are employed for rapid and cost-effective selection techniques (Shannon and Noble, 1990; Munns et al., 2002). Thirdly, the evaluation of the degree of salinity tolerance between and within species is likely to vary according to the criteria used for evaluation (Shannon, 1997), then developing an effective screening procedure may in itself be difficult (Flowers and Yeo, 1995). Last but not the least, evaluating salinity tolerance is made more complicated by the variation in salinity sensitivity during growth stages, i.e. the salt tolerance of the plants at one growth stage is not always correlated with tolerance at other stages (Greenway and Munns, 1980; Flowers, 2004; Foolad, 2004). Barley has been found to be tolerant to salinity at germination, sensitive at the seedling and early vegetative growth stages, then again tolerant at maturity (Epstein et al., 1980; Munns, 2002).

Various screening methods based on physiological traits have been practically and intensively employed in many plant breeding programs (Noble and Rogers, 1992; Shannon, 1997). These included germination percentage, degree of leaf injury, root length and plant height, shoot and root dry weight, shoot number, maintenance of flowering, seed and fruit set, canopy volume and quality, plant survival under salinity, tissue and specific accumulation of ions in different cell compartments (e.g. Na^+, K^+, Cl^-, Ca^{2+}), or their ratio (e.g. K^+/Na^+, Ca^{2+}/Na^+), and the production of certain metabolites or enzymes (e.g. proline, glycine betaine, sucrose, antioxidative enzymes) (Shannon, 1997; Munns et al., 2002; Munns and James, 2003; Foolad, 2004; Colmer et al., 2005). The use of physiological traits as screening tools has also been investigated for rice (Garcia et al., 1995) where it has proved successful in generating salt-tolerant lines (Gregorio et al., 2002). Thus, it may be also applicable to other crops (Cuartero et al., 1992; Ellis et al., 1997; Foolad, 1997; Munns et al., 2002). Despite this, there is still controversy about the use of a simple screening method for salt tolerance. Isla et al. (1997; 1998) reported that some physiological traits such as leaf ion concentrations, canopy temperature, stomatal conductance and grain ash content would not be useful screening tools for salinity tolerance in barley, and that grain yield remains the only reliable measure of identifying salt-tolerant barley. This section will review the most frequently used and recently developed screening methods for salt tolerance in barley.

3.1 Germination and emergence

High salt concentrations in the seed-planting layer of soil frequently lead to germination failure (Fowler, 1991), although as one of the most salt-tolerant crop species, some barley genotypes are reported to be able to germinate in seawater (i.e. 47 dS m^{-1}) (Mano and Takeda,

1995; 1997a). Salinity affects germination by multiple mechanisms including prevention of water uptake and exposure to ionic stress (Bewely and Black, 1982; Poljakoff-Mayber et al., 1994), reduction in enzyme hydrolysis and activities (Filho and Sodek, 1988; Guerrier, 1988), disturbance of the nitrogen metabolism (Yapsanis et al., 1994; Dell' Aquila and Spada, 1993), and imbalance of plant growth regulators (Khan and Rizvi, 1994).

In seed germination studies in Petri dishes under high salinity, Al-Karaki (2001) and Tajbakhsh et al. (2006) found that salt-tolerant barley maintained a much higher germination rate and shorter germination time compared to salt-sensitive barley. Germination percentage had a strong positive correlation with barley seed and straw yield (Thalji and Shalaldeh, 2007). Germination is a convenient test for screening a large number of genotypes, but little or no correlation has been found between genotypic differences in germination and later growth in salinity for many species (Kingsbury and Epstein, 1984; Ashraf and McNeilly, 1988; Mano and Takeda, 1997a; Almansouri et al., 2001; Shannon, 1997; Munns and James, 2003). For example, Donovan and Day (1969) found that a third out of 39 barley cultivars exceeded the germination of the best-known salt-tolerant cultivar California Mariout under high salinity. Emergence rate has also been reported as one of the criteria for salinity tolerance (Tajbakhsh et al., 2006) and it has been proposed to be a more practical screening criterion than germination rate (Shannon, 1997; Murillo-Amador et al., 2001).

3.2 Plant growth in response to salinity

Soil salinity reduces plant water uptake, leading to slower growth (Munns et al., 2006) along with a clear plant stunting as salt concentration increases (Wang and Nil, 2000). The components of this effect on plant growth are the reduction in plant height, fresh and dry weight of leaves, stems, and roots, yield and deterioration of the quality of the product (Kumar, 1995; Ali-Dinar et al., 1999; Chartzoulakis and Klapaki, 2000). One of the major causes of this growth reduction is inadequate photosynthesis owing to stomatal closure, inhibition of photosynthetic enzymes and limited CO_2 uptake (Zhu, 2001; Munns, 2002).

Plant water status changes rapidly and relative water content (RWC) is a convenient method of assessing plant water status. Jensen (1982) reported that the decrease in leaf osmotic potential was due partly to dehydration (58%) and partly to an increase in leaf solute content (42%). The turgor pressure was unaffected, but the RWC was significantly decreased by salinity in some wheat genotypes that accumulated high levels of leaf Na^+. At the same time, the low leaf Na^+ accumulating genotypes significantly reduced both turgor pressure and RWC after 5 days of salt treatment (James et al., 2002; Rivelli et al., 2002), suggesting a rather complex relationship between plant water status and ionic relations under saline conditions.

Plant growth components such as plant height, leaf and root elongation rate and biomass production under saline relative to non-saline conditions are frequently employed as simple and effective screening criteria (Kingsbury and Epstein, 1984; Munns and James, 2003). Plant height was used as an indicator of salinity tolerance in many species (Joshi and Nimbalkar, 1983; Forster, 2001; Houshmand et al., 2005). Seedlings of the salt-tolerant barley cultivar California Mariout (from which CM72 and Numar originated) showed no growth retardation at 400 mmol L^{-1} NaCl, while other genotypes were severely affected at lower salt concentrations (Epstein et al., 1980).

In early growth responses of wheat and maize to salinity, varietal differences in the elongation rate of leaves and roots are reported (Kingsbury et al., 1984; Mladenova, 1990). Aslam et al. (1993) and Moons et al. (1995) showed early varietal differences in rice and subsequent reductions in vegetative and reproductive yields in response to salinity. Moreover, early varietal differences in leaf growth responses to salinity were detected after only 3 days in two *Brassica* varieties whose leaf Na^+ and Cl^- levels were identical (He and Cramer, 1993).

In barley and wheat grown under salinity genotypic differences in the rate of leaf growth took a few weeks time to appear (Munns and James, 2003). However, within two weeks dead leaves became visible on the more sensitive genotype (Munns et al., 1995; Munns et al., 2000). Similarly, two maize cultivars with a 2-fold difference in Na^+ accumulation in leaves showed the same growth reduction. It was not until after eight weeks that a growth difference was clearly seen (Cramer et al., 1994; Fortmeier and Schubert, 1995). Thus, long-term experiments (several weeks to months) are necessary to assess genotypic differences in salt tolerance of cereal crops such as maize, wheat, barley, and rice using plant growth criteria (Kingsbury and Epstein, 1984; Aslam et al., 1993; Fortmeier and Schubert, 1995; Munns et al., 1995; Zhu et al., 2001).

3.3 Leaf element accumulation

3.3.1 Shoot/Foliar Na^+ content

Leaf Na^+ content analysis has the advantage of being directly related to the rate of Na^+ transport to the shoot, thus it is a specific index of Na^+ toxicity in the mesophyll. Na^+ exclusion is frequently used as a screening technique because genetic differences in exclusion are highly correlated with differences in salinity tolerance between durum and bread wheat (Francois et al., 1986; Gorham et al., 1987). In barley, Forster et al. (1994) showed that a salt-tolerant mutant was able to limit the amount of Na^+ uptake during salt treatment compared to its isogenic parent. Salt-tolerant barley varieties showed significantly lower Na^+ concentrations than those of sensitive varieties under saline conditions (Chen et al., 2005; Tajbakhsh et al., 2006). Extensive screening for salt tolerance in barley, based on Na^+ content, has been conducted in thousands of accessions from the world barley collections (Kingsbury and Epstein, 1984). Salt tolerance in barley varieties such as CM67 has been correlated with their ability to exclude Na^+ from the shoot (Jones and Storey, 1978; Royo and Aragues, 1993; Royo and Aragues, 1999). In addition, leaf Na^+ accumulation has been shown to relate to salt sensitivity in genotypes of rice, sorghum, and wheat (Yeo and Flowers, 1986; Munns, 2002; Munns et al., 2006; Krishnamurthy et al., 2007; Thalji and Shalaldeh, 2007). It is no exaggeration to say that this trait is the most popular among plant breeders working on improving salt tolerance in plants. There is one hurdle, however, which limits its applicability and reduces the prognostic value of such an approach. Shoot/foliar Na^+ analysis fails to take into account a plant's ability for Na^+ sequestration both at the cellular and tissue levels. Therefore, some salt-tolerant genotypes with efficient vacuolar compartmentation (enhanced Na^+/H^+ activity for example) may be simply missed when the selection is made based on shoot Na^+ content analysis.

3.3.2 Shoot/Foliar K^+ content

Potassium is the most abundant cation in higher plants and comprises up to 10% of the total plant dry weight (Marschner, 1995). In plants, K^+ plays central roles including osmoregulation, maintenance of turgor pressure, leaf and stomatal movement, enzyme activation, cell elongation, phloem solute transport, cation : anion balancing, control of membrane polarisation, cytoplasmic pH regulation, protein and starch synthesis, and energy conservation across membranes (Jones et al., 1979; Clarkson and Hanson, 1980; Kochian and Lucas, 1988; Maathuis et al., 1997; Leigh, 2001; Palmgren, 2001; Mäser et al., 2002a; Subbarao et al., 2003). The overall contribution of K^+ to the total solute potential, studied in over 200 plant species, varies from 66% to 90% (Wagner, 1982; Hsiao and Lauchli, 1986). The formation of the chloroplast structure, the translocation of assimilates and storage in the sink tissues are all processes critically depend on adequate tissue K^+ concentrations (Flowers and Lauchli, 1983). Finally, many metabolic processes and enzymatic reactions in the cytoplasm have a specific requirement for K^+ (Jones et al., 1979; Marschner, 1995).

K^+ levels in the soil solution range from 1 to 10 mmol L^{-1}, but intracellular K^+ levels in plants are maintained at 100–200 mmol L^{-1} (Kochian and Lucas, 1988). Cytoplasmic K^+ levels are well buffered against change by the large vacuolar pool of K^+, but high salt causes a decrease in cellular K^+ content and in the K^+/Na^+ ratio (Storey and Jones, 1978). Compared with barley, salt-tolerant wild *Hordeum* species maintain higher leaf K^+ under severe salt treatment (Garthwaite et al., 2005). Salt toxicity on seeds is reflected by decreases in seed K^+ concentration, and leakage of K^+ has been used as a measure of membrane damage from salinity (Nassery, 1979; Petruzzelli et al., 1992). Soil salinization (170 mmol L^{-1} NaCl) significantly decreases K^+ content in the shoots of salt-sensitive barley cultivars, but the K^+ content in the most tolerant one increased (Leonova et al., 2005). Salt treatment also reduces potassium activity in the barley epidermal leaf cell vacuoles from 224 to 47 mmol L^{-1} and in the cytosol from 68 to 15 mmol L^{-1}, while the corresponding changes in the mesophyll were from 235 to 150 mmol L^{-1} (vacuole) and 79 to 64 mmol L^{-1} (cytosol) (Cuin et al., 2003).

Plants that are more tolerant to salt have a greater ability to maintain high levels of K^+. This has been found in crops screened and bred for salinity tolerance, as well as in wild relatives of certain crop species (Colmer et al., 1995; Dubcovsky et al., 1996; Flowers and Hajibagheri, 2001; Zhu et al., 2001). For example, Thalji and Shalaldeh (2007) reported that, for barley and wheat, K^+ content at the three-leaf stage showed a strong positive correlation with seed yield that is the ultimate criterion of salinity resistance. Chen et al. (2007a) also found that K^+ flux from the root in response to NaCl treatment (the ability to maintain K^+ under salt stress) was highly inversely correlated with relative grain yield, shoot biomass, plant height, net CO_2 assimilation, survival rate and 1,000-grain weight measured in glasshouse experiments after 4–5 months of salinity treatment.

3.3.3 K^+/Na^+ and Ca^{2+}/Na^+ ratios

The capacity of plants to maintain a high cytosolic K^+/Na^+ ratio is one of the key determinants of plant salt tolerance (Serrano et al., 1999; Amtmann et al., 2004). Under typical physiological conditions, plants contain about 100 mmol L^{-1} K^+ and maintain a high K^+/Na^+ ratio in their cytosol (Binzel et al., 1988), rarely tolerating cytosolic Na^+ levels above 20 mmol L^{-1} (Blumwald et al., 2000). *In vitro* activities of enzymes extracted from the halophytes pop saltbush or seablite (*Suaeda maritima* L.), and even enzymes from the pink salt-lake alga (*Dunaliella parva*) that tolerates 10-fold-seawater salinity, were just as sensitive to NaCl as were those of beans or peas (Greenway and Osmond, 1972; Flowers et al., 1977). In contrast to durum wheat, barley is more efficient on Na^+ and K^+ partitioning in cellular and subcellular compartments, leading to the preservation of a higher cytoplasmic K^+/Na^+ ratio at high leaf Na^+ concentrations (Munns et al., 2006). This has been confirmed by electrophysiological analysis of root cation channels. All major K^+ influx channels exhibited higher K^+/Na^+ selectivity in salt cress (*Thellungiella halophila*) than in *Arabidopsis* (Volkov et al., 2004). Rivelli et al. (2002) showed that in low-Na^+ wheat genotypes, osmotic adjustment was enabled by a higher K^+ content, Na^+ exclusion being associated with maintenance of higher K^+ levels. Cytoplasmic Na^+ concentration in the salt-sensitive barley cultivar was almost 1.4 times greater than that of the salt-tolerant one, with the latter also having a higher K^+ concentration. It is evident that a salt-tolerant cultivar has a higher Na^+ concentration in its cytoplasm than a salt-resistant variety (Flowers and Hajibagheri, 2001).

The higher Na^+ uptake and lower K^+ content exhibited by the salt-sensitive barley have been contrasted with the lower Na^+ and higher K^+ content in the salt-tolerant cultivar California Mariout and its derivative CM67 (Jones and Storey, 1978; Gorham et al., 1994). Similar results have also been reported by Al-Karaki (2001). Tissue K^+/Na^+ ratio has been used successfully in the selection of many crops for salinity tolerance (Janardhan et al., 1979; Chhipa and Lal, 1985; Dvořák et al., 1994; Asch et al., 2000; Tajbakhsh et al., 2006; Thalji and Shalaldeh, 2007). Thalji and Shalaldeh (2007) suggested that the K^+/Na^+ ratio in barley and wheat can be used as a selection criterion for salt tolerance because it is highly correlated with biomass, seed and straw yields. The Ca^{2+}/Na^+ ratio also appears to be a more reliable indicator of salt stress than Na^+ content alone (Ben-Hayyim et al., 1987; Krishnamurthy et al., 2007). Indeed, the maintenance of higher K^+/Na^+ and Ca^{2+}/Na^+ ratios in young growing tissues of barley does appear to be an important mechanism contributing to improved salt tolerance (Wei et al., 2003).

Controversially, bulk leaf ion concentrations of Na^+, K^+, Ca^{2+}, and the K^+/Na^+ and Ca^{2+}/Na^+ ratios were not judged to be the cause of the differences in grain yield that were observed in wheat and barley cultivars. Thus, it was suggested that these ratios could not be used in screening for salt tolerance (Rawson et al., 1988; Isla et al., 1997). Moreover, Fricke et al. (1996) reported that studies using the bulk leaf ion concentrations may be misleading because they do not detect potential ion exclusion mechanisms by the cytoplasm. On the other hand, they showed that Na^+ and Cl^- in different leaf compartments could be relevant to barley salt tolerance. Nevertheless, bulk leaf ion concentrations of Na^+, K^+, Ca^{2+}, and the K^+/Na^+ and Ca^{2+}/Na^+ ratios are considered reliable indicators of salt tolerance and they are, therefore, some of the most frequently used physiological screening methods (Munns and James, 2003) and phenotyping indices in molecular marker studies (Dubcovsky et al., 1996; Koyama et al., 2001; Lin et al., 2004). In an experiment with 69 randomly selected barley varieties, Chen et al. (2007a) found that 62 out of 69 cultivars followed an inverse relationship between K^+ efflux (release of K^+ under salt stress) and salt tolerance. In a few cultivars, however, high salt tolerance (measured as grain yield at harvest) was observed for plants showing only modest ability to retain K^+ in the root cells. Tissue elemental analysis showed that these plants had a much better ability to prevent Na^+ accumulation in plant leaves and, thus, to maintain a higher K^+/Na^+ ratio. They concluded that a plant's ability to maintain a high K^+/Na^+ ratio (by either retention of K^+ or preventing Na^+ from accumulating in leaves) is a key feature for salt tolerance in barley.

3.4 Photosynthetic parameters

3.4.1 Photosynthetic rate

Under mild salt treatment (150 mmol L^{-1} NaCl), the photosynthetic rate (Pn) of barley plants was only slightly affected by the NaCl treatment (Fricke et al., 1996). However, Pn, g_s, and C_i of a number of barley genotypes significantly decreased under exposure to high salinity (Jiang et al., 2006). Furthermore, these responses differed greatly between salt-tolerant and -sensitive genotypes (Huang et al., 2006). Tajbakhsh et al. (2006) reported that the Pn of barley decreased with the addition of salt and the cultivars showed significant differences in the rate of decline under greater salinity stress. The impact of salinity on the CO_2 assimilation also differed substantially among barley cultivars, with tolerant genotypes maintaining net CO_2 assimilation 3-fold to 4-fold higher than the sensitive ones (Chen et al., 2007a). Munns and James (2003) and James et al. (2006) reported that the maintenance of the photosynthetic capacity parameters of barley compared to durum wheat at higher leaf Na^+ levels were associated with the maintenance of higher K^+, lower Na^+ and a resultant higher K^+/Na^+ ratio in the cytoplasm of the mesophyll cells in barley. If the major limitation to photosynthesis is the stomatal conductance, then this parameter may be an effective way of selecting wheat genotypes for cultivation in saline soils (Rivelli et al., 2002; Munns and James, 2003). However, screening methods based on photosynthesis are not suitable for handling a large number of genotypes, except stomatal conductance measured by viscous flow porometry (Rebetzke et al., 2000; James et al., 2002).

3.4.2 Chlorophyll fluorescence

Chlorophyll fluorescence was employed for measuring salinity-induced inhibition of PSII (Abadia et al., 1999; Fedina et al., 2002) and thus in screening of barley genotypes for salinity tolerance under controlled conditions (Belkhodja et al., 1994; 1999). A two-week exposure to 20 dS m^{-1} saline conditions significantly reduced the number of chlorophyll fluorescence parameters such as the Φ_{PSII} and *qp* in barley (Jiang et al., 2006). In rice (Yamamoto et al., 2004) and wheat (Muranaka et al., 2002) seedlings exposed to 100 mmol L^{-1} NaCl, Φ_{PSII} was markedly decreased in both salt-tolerant and -sensitive genotypes, but the photosynthetic activity was maintained in the salt-tolerant lines (Muranaka et al., 2002). Under high salinity, both the slow-down of the electron transport at the PSII in sunflower (Rivelli et al., 2002) and the reduced photochemical efficiency of olive (Loreto et al., 2003) have been reported, although neither was associated with the photosynthetic activity reduction.

F_v/F_m values have been widely used as a non-destructive and non-invasive tool to determine effects of environmental stresses on the photosynthetic apparatus (Maxwell and Johnson, 2000; Shabala, 2002; Sayed, 2003). However, the salinity showed no apparent effect of salt stress on the maximal quantum efficiency of PSII (F_v/F_m). Regardless of the severity of the salt stress, the F_v/F_m values remained above 0.8, indicating optimal functioning of PSII (Chen et al., 2005), thus even under severe salt stress conditions the leaf photochemistry in barley is well protected. In maize, F_v/F_m was also not significantly affected by NaCl treatment (Shabala et al., 1998), and F_v/F_m, Φ_{PSII}, *qp* and non-photochemical quenching (NPQ) showed little differences between wheat genotypes in response to 150 mmol L^{-1} NaCl, indicating that the efficiency of PSII photochemistry was not affected by salinity (Rivelli et al., 2002). Thus, it is unlikely that chlorophyll fluorescence measurements can be used as a reliable tool in breeding programs aimed at improving salinity tolerance in plants. Although measuring gas exchange and chlorophyll fluorescence are not suitable for screening a large number of genotypes for salinity tolerance, they are still valuable to investigate the effects of salinity on photosynthetic mechanisms in a few germplasm.

3.5 Grain/biomass yield

Grain yield is the ultimate criterion of field performance and the main aim of the entire breeding process. It is, therefore, obvious that the grain yield of salt-tolerant barley cultivars is less affected by NaCl than that of the salt-sensitive cultivars (Chauhan et al., 1980; Isla et al., 1998; Flowers, 2004; Leonova et al., 2005). A salt stress treatment significantly reduced grain yield, shoot biomass and 1,000-grain weight of the barley plants. The impact of salinity differed substantially between barley cultivars, i.e. salt-sensitive genotypes gave zero grain yield under saline conditions whilst salt-tolerant genotypes yielded 25%–30% of their control (non-saline) values (Chen et al., 2007a). This reduction in grain yield and shoot biomass was most likely a consequence of reduced CO_2 assimilation under saline conditions (Chen et al., 2007a). The forage yield of the Omani Batini barley landrace at tillering stage under 10 dS m^{-1} and 20 dS m^{-1} can be predicted with high and moderate accuracy by using the forage yield of this genotype grown under 0.85 dS m^{-1} (Jaradat et al., 2004). Interestingly, in a field testing of 124 barley genotypes under ten salinity treatments over five consecutive years, Royo and Aragües (1999) found that the most productive genotypes were not necessarily the least salt-tolerant, so it might be useful to select for the most productive barley genotypes under medium and high saline conditions. It is important to note that controlled-environment chambers or glasshouses cannot provide adequate conditions (space, light, water and pot size) required to predict maximum field yield, hence, the field evaluation of the yield potential under salinity is critical in the breeding for salt-tolerant crops. However, field experiments for yield performance may be more appropriate at the final stages of breeding programs rather than at the initial stages when screening for salt tolerance is best done under controlled environments (Shannon, 1997; Zeng et al., 2002).

3.6 Survival

Survival under high salinity is also a convenient test for salinity tolerance (Kingsbury and

Epstein, 1984; Sayed, 1985; Tal, 1985). The rate (percentage) of survival in saline conditions has been used to evaluate the salt tolerance of 24 barley genotypes and eight wild *Hordeum* in the glasshouse and controlled environment cabinet (Flowers and Hajibagheri, 2001; Garthwaite et al., 2005). In an experiment with 69 barley varieties, the survival rate was greatly limited by severe salinity. The survival rate was correlated strongly with plant salt tolerance estimated as grain yield (Chen et al., 2007a). Furthermore, survival under salinity has been employed as a physiological indicator and its QTLs were detected in rice (Lin et al., 2004). The physiological and genetic factors that contribute to the growth of crops at very high salt concentrations were found to be proportionally related to survival more than to high yields, though survival is not of interest to the farmers (Shannon and Noble, 1990). In spite of the considerable genetic diversity existing among 5,000 hexaploid and tetraploid wheat lines, only a marginal correlation was found between survival under high salinity and performance in the field (Sayed, 1985). Survival at high NaCl as a selection criterion is rapid and simple but it does not necessarily imply healthy growth and carries the risk of selecting against productivity.

3.7 Accumulation of compatible solutes

Accumulation of compatible solutes is a typical plant response to salinity exposure (Yancey et al., 1982; Hare et al., 1998). Compatible solute accumulation has long been emphasized as a selection criterion in traditional crop breeding programs (Morgan, 1984; Ludlow and Muchow, 1990). Recent progress in molecular biology has made this approach central to molecular breeding programs, largely due to the fact that the osmolyte accumulation is often controlled by only a single gene (Serraj and Sinclair, 2002). At 300 mmol L^{-1} NaCl, glycine betaine and proline together contribute almost 15% to the osmotic potential in leaves of the halophytic sea barley grass, compared to only 8% in barley (Garthwaite et al., 2005). Glycine betaine and proline concentrations in the flag leaf of plants exposed to 200 mmol L^{-1} NaCl were much higher in the salt-tolerant sea barley grass than those in the salt-sensitive bread wheat, and salinity tolerance was found expressed in their amphiploid (Islam et al., 2007). In barley, the impact of H_2O_2 (one of the components of salt stress) on the K^+ flux (a measure of stress "severity") and the mitigating effects of glycine betaine and proline on the NaCl-induced K^+ efflux were found to be significantly higher in salt-sensitive barley genotypes (Chen et al., 2007a). The salt-sensitive cultivars showed a 2-fold higher accumulation of leaf and root proline and leaf glycine betaine. The total amino acid content was less affected by salinity in salt-tolerant cultivars. In salt-sensitive genotypes, glycine betaine and proline contributed substantially to cell osmolality compensating for reduced cytosolic K^+. They concluded that the hyperaccumulation of known major compatible solutes in barley does not appear to play a major role in salt tolerance but rather it may be a symptom of salt-susceptibility (Chen et al., 2007a). Due to the controversy over whether high accumulation of compatible solutes is actually beneficial for salinity tolerance in glycophytes, the use of salt-induced compatible solute accumulation is not validated as a potential screening tool for salinity tolerance. It also differs among plant species.

3.8 The microelectrode ion flux estimation (MIFE) technique

The proposal to use ion concentrations or electrochemical potentials measured outside plant tissues to estimate the net ion fluxes of tissues came from B. Lucas (Lucas and Kochian, 1986). Over the past twenty years, the microelectrode ion flux estimation (MIFE) technique has been employed in a broad range of research areas in plant sciences (Newman, 2001; Shabala, 2002; 2006), including the response of plants to salinity (Shabala et al., 1998; 2003; 2005), waterlogging (Pang et al., 2006a; 2006b), Al^{3+} toxicity (Wherrett et al., 2005), Ca^{2+} and Mg^{2+} deficiency (Shabala et al., 2003; Shabala and Hariadi, 2005), high and low temperatures (Shabala, 1996; 1997), rhythmic patterns of nutrient acquisition (Shabala and Knowles, 2002; Shabala et al., 2006b), plant response to blue light (Babourina et al., 2002), fluctuations of light intensity (Živanović et al., 2005), and plant ion transporter studies, i.e. combined with patch clamp and other advanced techniques for detailed ion transporter studies (Tyerman et al., 2001; Shabala and Lew, 2002; Demidchik et al., 2003; Shabala et al., 2006a).

The use of the MIFE technique has advanced research into plant salinity tolerance (Shabala, 2006; Chen et al., 2005; 2007a; 2007b). For example, the bioelectric response measured by MIFE was reported to be a sensitive indicator of NaCl stress in maize leaves (Shabala et al., 1998). Indeed, it has been shown that NaCl stress results in a significant net K^+ efflux (prevented by 10 mmol L^{-1} Ca^{2+}) from the leaf mesophyll of broad bean (*Vicia faba* L.). In contrast, plants showed a net K^+ uptake in response to an isotonic mannitol application. These differences reflect the involvement of both the ionic and osmotic components of the salinity stress (Shabala, 2000). In suspension cells from wheat, fusicoccin prevents NaCl-induced K^+ loss from the cell by direct activation of H^+-ATPases and other metabolic changes crucial for the plant's adaptation to high salinity (Babourina et al., 2000). It was found that there were no effects of NaCl on the net Ca^{2+} flux in protoplasts from the broad bean mesophyll indicating that the large transient NaCl-induced Ca^{2+} efflux from tissue originates from ion exchange in the cell wall (Shabala and Newman, 2000). Shabala et al. (2003) reported that NaCl causes a rapid and prolonged efflux of H^+, K^+, and NH_4^+ from the root epidermis with a more positive plasma membrane potential (E_m). The relative divalent cation efficiency on stimulating barley Na^+ efflux and K^+ uptake was shown to be $Ba^{2+}>Zn^{2+}=Ca^{2+}>Mg^{2+}$ (Shabala et al., 2005). In addition to their ability to block non-selective cation channels (NSCCs), divalent cations also control the activity of K^+ transporters to maintain the high K^+/Na^+ ratio required for optimal leaf photosynthesis. It has been proposed that the compatible solutes prevent NaCl-induced K^+ efflux from the barley roots through enhancing the activity of H^+-ATPase, thereby controlling the depolarisation-activated potassium channels (DAPCs) and creating the electrochemical gradient necessary for secondary ion transport processes (Cuin and Shabala, 2005). Moreover, Cuin and Shabala (2007) found that a large number of amino acids cause significant mitigation of the NaCl-induced K^+ efflux from barley root epidermis, thus suggesting that free amino acids might also contribute to plant adaptation to salinity by regulating the K^+ transport across the plasma membrane. A very strong negative correlation was found between the magnitude of K^+ efflux from the root and the salt tolerance of a particular barley cultivar (Chen et al., 2005). In intact 3-day old seedlings following 40 min pretreatment with 80 mmol L^{-1} NaCl, the K^+ efflux from the mature root-zone was found to be a reliable screening indicator for salinity tolerance in barley (Chen et al., 2005). Further experiments showed that the MIFE technique can distinguish contrasting barley varieties differing substantially in salt tolerance (Chen et al., 2007a).

In summary, various physiological, electrophysiological, genetic, and biochemical methods can be employed to unravel the mechanisms underlying the differential responses to salinity between salt-tolerant and -sensitive barley genotypes. Also, a newly developed screening method based on NaCl-induced ion flux measurement (MIFE technique) can facilitate the rapid screening of plant species and cultivars for their salinity tolerance.

4. BARLEY GERMPLASM FOR SALINITY TOLERANCE

The gene pool of the wild relatives of barley may represent a valuable source of new loci for salt tolerance. Mano and Takeda (1998) evaluated the salt tolerance of 340 accessions of *Hordeum*, consisting of 41 brittle-rachis forms of *Hordeum vulgare* L. subsp. *vulgare* (*H. agriocrithon)* accessions, 154 *H. vulgare* L. subsp. *spontaneum* (*H. spontaneum)* accessions, and 145 accessions of ten other species or subspecies of wild *Hordeum*. They found that the levels of salt tolerance for seed germination in the wild *Hordeum* species were generally lower than those in cultivated barley and the NaCl tolerance level of the different species was as follows: *H. agriocrithon* > *H. spontaneum* > other wild *Hordeum* species. The high salt tolerance level of *H. agriocrithon* probably originated from the natural hybridization between *H. spontaneum* and cultivated barley (Briggs, 1978). In addition, based on the leaf injury index at the seedling stage, the levels of salt tolerance in wild *Hordeum* species were generally higher than those found in cultivated barley. Most wild *Hordeum* species showed high NaCl tolerance at the seedling stage and were considered good sources of germplasm in the barley breeding for salt tolerance.

The natural populations of *H. spontaneum* showed rich genetic variation in tolerance to physiological stresses such as salinity. This was highlighted by the large variation both between and within populations observed in the extreme desert station at Sede Boqer (Nevo et al., 1984). Subsequently, populations and genotypes of the wild barley *H. spontaneum* and the wild emmer wheat *Triticum dicoccoides* were tested for salt tolerance under controlled conditions using 150 mmol L^{-1}, 250 mmol L^{-1} and 350 mmol L^{-1} NaCl (equivalent to 25%, 40% and 60% sea water) with non-saline controls (Nevo, 1992). They found wide differences in high salt tolerance within and between populations of *H. spontaneum* and *T. dicoccoides* and identified wild barley genotypes with high tolerance in populations from the coastal mediterranean areas (e.g. Caesarea), the Jordan Valley (Mehola) and the northern Negev Desert (Makhtesh Gadol). The salt-tolerant lines could spike normally under 350 mmol L^{-1} NaCl. Natural populations of the wild barley can be found southwards to the northern Negev Desert, yet they only grow there in dry watercourses, thereby effectively utilizing the extra run-off water. Under extremely harsh ecological conditions such as an arid water regime of 130 mm and highly saline soil conditions, some populations and superior genotypes have proved relatively outstanding performance in both growth and yield. Thus, the genetic resources of the barley progenitors display remarkable morphological and physiological performances, which are of potential economic importance for improving crop salt tolerance.

There is variation among barley varieties for salinity tolerance. Almost 5,000 barley lines were screened in Syria for germination, seedling vigour and adult plant vigour (Srivastava and Jana, 1984). Promising lines were identified but the experiments did not result in the release of new varieties with salinity tolerance. Ramagopal (1988) found differences in the protein synthesis during seed germination between salt-tolerant barley genotype California Mariout 72 (CM72) and the salt-sensitive genotype Prato. Salinity stress induced both quantitative and qualitative changes in the expression of some proteins *in vivo.* Around 8% of the nearly 400 resolved tissue proteins were affected in this way. Some of the proteins in this category were specific to each genotype. About 1% of the total showed qualitative changes; these proteins were expressed only during salinity stress. In roots, two proteins were detected in CM72 and five in Prato. In shoots, four proteins were found only in Prato and these were similar to those induced in roots. The four new proteins in germinating embryos were apparently induced only in CM72. It is highlighted that ontogeny plays an important role in the expression of tissue-specific proteins during salinity stress in the salt-tolerant and -sensitive barley genotypes.

5. DEVELOPMENT OF SALINITY TOLERANT GERMPLASM THROUGH *IN VITRO* CULTURE

Ye et al. (1987) demonstrated that *in vitro* anther culture in a salt-containing medium was an effective approach to develop salt-tolerant barley lines. Anthers of two six-rowed barley parent cultivars Diamond (salt-sensitive at germination) and Men Yuan Liang Lan (salt-tolerant), and their F_1 from reciprocal crosses were cultured in a liquid medium containing 0%, 0.4%, 0.6%, and 0.8% Na_2SO_4. Among 37 progenies from the F_1 pollen in the salt (Na_2SO_4)-free medium, it was found that 11 were as sensitive as Diamond, 12 were intermediate to the two parents, 7 were equal to the salt-tolerant parent and 7 were more tolerant to Na_2SO_4 than Men Yuan Liang Lan. Whilst in the high salt-containing medium no progeny from the F_1 pollen was as susceptible as the susceptible parent; 2 were intermediate, 2 were equal to the salt-tolerant parent and 2 were more tolerant than the salt-tolerant parent. The results demonstrated that culturing anthers in the Na_2SO_4 -containing medium effectively eliminated salt-sensitive progenies. All 16 microspore-derived lines of Diamond were as susceptible as Diamond to Na_2SO_4 whereas 5 microspore-derived lines from Men Yuan Liang Lan were as tolerant as Men Yuan Liang Lan to Na_2SO_4. The results indicated that the lines exhibited elevated levels of salt tolerance probably resulting from the recombination of genes rather than from spontaneous mutation.

6. DEVELOPMENT OF SALINITY TOLERANT GERMPLASM THROUGH MUTATION

Using EMS induction and salt-induced tissue culture in combination with irradiation, four salt-tolerant mutants from *Hordeum brevisubulatium* were identified (Lu et al., 2002). The most successful example using mutation for barley salinity tolerance is the cultivar Golden Promise, a γ-ray induced semi-dwarf mutant of the cultivar 'Maythorpe', although salinity tolerance was not the initial purpose. Forster (2001) reviewed the research made at the Scottish Crop Research Institute (SCRI) on the effects of semi-dwarf genes on salt tolerance. This work was initiated in 1993 with the fortuitous and unexpected result that the cultivar Golden Promise showed considerable tolerance to salt. The parent and mutant showed significant differences in their responses to salt stress. The positive and pleiotropic effects of the mutant gene *Gpert* were found to be effective in a number of genetic backgrounds. The *Gpert* mutation was allelic to the *ari-e* mutants in barley. The *ari-e* mutants were salt-tested and found to show the same positive responses to salt stress as Golden Promise, supporting the allelism tests, and consequently the *Gpert* symbol was changed to *ari-e.GP*. The semi-dwarf mutant *sdw1* and the *erectoides* semi-dwarf mutant *ert-k32* were also tested for their effects on salt tolerance, but they did not show any positive effects. Salt tolerance was, therefore, not a general phenomenon of semi-dwarf stature but specific to the mutations at the *ari-e* locus in these lines. The barley varieties Dash and Hindmarsh with the *ari-e.GP* gene showed excellent yield potential and adaptation to the

Australian environment. As there are closely linked molecular markers available for the *ari-e.GP* gene, it should be a good target for MAS to improve barley salinity tolerance. Wei et al. (2001) found a gene encoding the barley vacuolar ATPase subunit B (*BSVAP*), which was differentially expressed between the near isogenic barley cultivars, Golden Promise and Maythorpe. The gene was inducible under long-term salinity stress in the salt-sensitive cultivar Maythorpe, but less so in the relatively salt-tolerant Golden Promise and was highly expressed under control conditions in Maythorpe. It was concluded that the short-term down-regulation of *BSVAP* under high salinity was an important mechanism contributing to the salt tolerance of Golden Promise. This gene could be the target for genetic modification or MAS for improvement of salinity tolerance. Dietz et al. (2001) suggested that the ability to respond to salinity stress with changes in the gene expression of the vacuolar ATPase might be a pre-requisite and a characteristic of salt tolerance in plants.

7. MARKER-ASSISTED SELECTION (MAS) FOR SALINITY TOLERANCE

7.1 Barley molecular marker techniques

Successful implementation of MAS in a breeding program consists of nine key steps as follows: characterizing the genetics of traits, choosing the parents, constructing the genetic population, multiplying the seeds, map construction, phenotyping the mapped population, identifying genes/QTLs for the target traits, validating putative markers and implementing molecular markers in the breeding programs. Map construction is the most expensive and time-consuming step and it usually demands more than 40% of the total cost for MAS (Barr et al., 2000). Developments of the marker technologies in the last few years have significantly reduced the cost for map construction. Since the development of the first generation of restriction fragment length polymorphism (RFLP) barley molecular maps (Graner et al., 1991; Kleinhofs et al., 1993), Polymerase chain reaction (PCR)-based molecular markers have become the dominant marker type for map construction and MAS (Varshney et al., 2004). Among the different types of molecular markers available for barley, microsatellite or simple sequence repeats (SSRs) have become the most commonly-used markers for genetic diversity study, map construction and MAS in the breeding programs. The SSR markers are abundant in genomes, multi-allelic, highly informative, and co-dominantly inherited. In addition, SSR markers are of easy detection and amenable to high-throughput analysis. Several thousands of barley SSR markers have been developed from different research groups including SCRI (Bmac, Bmag, EBmac, EBmag, HVGeneName, scsssr), IPK (GBM, GBMS), WUR (GBM), Virginia Polytechnic Institute (HVM), and MPI for Plant Breeding (HVGeneName). These SSR markers have been mapped in various mapping populations.

Diversity arrays technology (DArT) is a new high-throughput molecular marker system, which belongs to a hybridization-based molecular marker method deployed on a microarray platform. The Technology was developed and commercialized by Triticarte P/L (http://www.diversityarrays.com). DArT is genome sequence-independent and allows high-throughput screening of hundreds of molecular markers simultaneously. It is especially suitable for generating genome-wide markers for mapping genetic linkages and identifying markers closely linked to genes or QTLs for MAS applications. Barley was the first species for which DArT markers were made available. Approximately 2.3 million data points for 4,000 lines have been generated for the barley breeders and researchers (Wenzl et al., 2006). More than 2,000 DArT markers have been mapped in various barley populations and integrated into a consensus map with RFLP, SSR and STS markers (Wenzl et al., 2006). The consensus map comprised 2,935 loci (2,085 DArT, 850 other loci) and spanned 1,161 cm. It contained a total of 1,629 'bins' (unique loci), with an average inter-bin distance of (0.7 ± 1.0) cm (median = 0.3 cm). More than 98% of the map could be covered with a single DArT assay.

Single Nucleotide Polymorphisms (SNPs) are the most abundant form of DNA polymorphism and now their applications predominate in the modern genetic analysis. Over 17,000 candidate SNPs have been identified from mining 350,000 public barley expressed sequences (Appleby et al., 2008). However, this number should be considered with caution as the greatest challenge of *in silico* SNP discovery is the differentiation between true polymorphisms and the sequence error. In another study, more than 2,000 genome-wide barley SNPs were developed by re-sequencing unigene fragments from eight diverse accessions (Rostoks et al., 2005). The average genome-wide SNP frequency observed in 877 unigenes was 1 SNP per 200 bp DNA sequence. However, the SNP frequency was highly variable with the least number of SNP and SNP haplotypes observed within the European cultivated germplasm reflecting effects of breeding history on the genetic diversity. More than 300 SNP loci were mapped genetically in three experimental mapping populations, which allowed the construction of an integrated SNP map incorporating a large number of RFLP, AFLP and SSR markers (1,237 loci in total). Recently, a total of 12,615 SNPs were identified in 3509 EST-derived contigs using 36 pairwise comparisons between nine barley genotypes (Bhat et al., 2007). A linear relationship was observed between the number of SNPs and the product of the number of ESTs for each pair of genotypes. An additional ~ 800 SNPs from genomic amplicon sequences were also used. An Illumina oligonucleotide pool assay (OPA) was designed for high-throughput SNP genotyping, representing 1,524 different barley unigenes with one SNP each. In summary, SSR, DArT and

SNP techniques provide efficient molecular tools for map construction, QTL identification and MAS for salinity tolerance in barley.

7.2 Mapping the genes/QTLs for salinity tolerance

A number of QTLs affecting salinity tolerance were detected on all chromosomes in *Hordeum spontaneum* and *Hordeum vulgare* and the crosses of the two species. In cultivated barley, Mano and Takeda (1997b) identified QTLs controlling salt tolerance at germination and the seedling stage in barley by interval mapping analysis using marker information from two doubled haploid (DH) populations derived from the crosses Steptoe × Morex and Harrington × TR306. The results revealed that the QTLs for salt tolerance at germination were located on the chromosomes 4H, 6H, and 5H of the DH lines of Steptoe × Morex, and on the chromosomes 1H and 5H of the DH lines of Harrington × TR306. In both DH populations, the most effective QTLs were found at different loci on the chromosome 5H. Genetic linkage between salt tolerance at germination and ABA response was found from QTL mapping. The QTLs for the most effective ABA response at germination were located very close to those for salt tolerance on the chromosome 5H in both crosses. The QTLs for salt tolerance at the seedling stage were located on the chromosomes 2H, 1H, 6H, and 5H in the DH lines of Steptoe × Morex, and on the chromosome 5H in the DH lines of Harrington × TR 306. Their positions were different from those of QTLs controlling salt tolerance at germination, indicating that salt tolerance at germination and at the seedling stage were controlled by different loci. QTLs were also identified for physiological traits hypothesized to be associated with the salinity tolerance (Ellis et al., 1997). However, it was found that they are not useful indicators for screening barley for grain yield under saline conditions (Isla et al., 1998). In a cross of Derkado × B83-12/21/5, the largest individual effects to salt tolerance were associated with the chromosomal regions around the two dwarfing genes *sdw1* (3H) and *ari-e.GP* (5H). The *sdw1* gene resulted in an overall yield increase, but it was only detected as a secondary QTL (Ellis et al., 2002). In a mapping population study of winter barley Angora × W704/137, germination under salt stress was related to QTLs on linkage groups 3H and 5H confirming previous results. A QTL related to sodium uptake not identified from the germination test was found on the chromosome 1H in the field trial (yield at maturity) whereas another QTL responding to salt stress on the chromosome 3H was found in both development stages. This QTL was related to yield reduction in the later development stages, probably indicating a gene location related to the translocation of carbohydrates (Weidner et al., 2007). QTLs controlling salt tolerance at later growth stages of barley were identified by interval mapping analysis using marker information from doubled haploid (DH) populations derived from the cross CM72 and Gairdner. Tillers per plant, plant height, spikes per line, spikes per plant, shoot dry weight, grains per plant, grain yield, shoot Na^+, K^+ content and Na^+/K^+ ratio were examined under salinity and no salinity (control) stress conditions. Totally 31 QTLs were detected, from which 16 under the control and 15 under the salt stress conditions. All these QTLs are located on 7 chromosomes. Moreover, QTLs affecting different characters or the same character under different treatments were clustered on the chromosomes. Interestingly, four QTLs characterized by pleiotropic effect were detected in the region of bPb-1278–bPb-9504 on the chromosome 4; these QTLs are associated with tillers per plant, spikes per line, spikes per plant and shoot dry weight, respectively. Moreover, the additive effect is mainly attributed to the salt-tolerant parent CM72.

However, it has been reported that such QTLs were dependent on the conditions under which the plants were grown (Foolad, 1999; Monforte et al., 1997). In addition, it was suggested that the QTLs associated with salt tolerance vary with the developmental stage of the plant. A further limitation to the use of QTLs in plant breeding is the fact that the QTLs may be specific to particular crosses. Therefore, it may be argued that whilst markers will be of value in using elite lines from the mapping population in backcrossing, the result cautions against any expectation of general applicability of markers for physiological traits. Although there is no successful report for using MAS to improve salinity tolerance, the identified loci/QTLs may provide a basis to integrate MAS with conventional approaches for the improvement of salinity tolerance in barley.

8. IDENTIFICATION OF CANDIDATE GENES FOR SALINITY TOLERANCE

Plants respond to high salinity conditions by adjusting their physiological and metabolic processes (Rhode and Hanson, 1993). There are many genes for maintaining ion homeostasis and metabolism including the synthesis of compatible solutes (osmo-protectants), stress proteins for cell rescue and defense, proteins for signal transduction, components of protein synthesis and others affecting morphology (Bohnert and Jensen, 1996; Kasuga et al., 1999; Kawasaki et al., 2001).

Candidate gene approaches have been used to study salinity tolerance. Munns (2005) and Colmer et al. (2005) reviewed the potential genes involved in salt tolerance. There are four categories of genes, which control traits that might be expected to increase salinity tolerance as follows: (i) proteins that control the transport of ions across membranes; (ii) enzymes that synthesise organic solutes that contribute to the control of plant water relations or functions; (iii) proteins involved in protecting against oxidative damage and other consequences of osmotic stress; and (iv) genes encoding components of signal transduction pathways.

The development of the microarray technology provides an opportunity to identify a large number of genes involved in salinity tolerance. The barley cultivar Morex was used for transcriptional profiling during salinity stress using a microarray containing 22,750 probe sets (Walia et al., 2006). The experiment was designed to target the early responsive genes to salinity stress at the seedling stage. A number of probe sets were found up-regulated and down-regulated in response to salinity. A prominent feature of the response to salinity was the induction of genes involved in jasmonic acid biosynthesis and genes known to respond to jasmonic acid treatment. A large number of abiotic stresses (heat, drought and low temperature) related genes were also found to be responsive to salinity stress. Osmoprotection was found to be an early response of barley under salinity stress. Unfortunately, the results of two reports characterizing gene expression of barley under salinity stress found very few genes in common (Ozturk et al., 2002; Ueda et al., 2004). The gene encoding pyrroline-5-carboxylate synthetase (P5CS) is the only gene found to be commonly up-regulated in all three studies.

A differential display using randomly amplified polymorphic DNA (RAPD) primers isolated a salt-inducible gene encoding a putative methionine synthase in barley leaves (Muramoto, 1999; Shi et al., 2001). The expression of this gene named as *HvMS* (*Hordeum vulgare* methionine synthase) is induced in barley leaves within 1 h under salt stress. This is one of the early responsive genes in barley leaves. It was assumed that the salt-inducible *HvMS* could play an important role as a member of this cycle for salt tolerance in the barley plants (Shi et al., 2001). Eckermann et al. (2000) reported that the expression of *methionine synthase (MS)* gene is induced under salt stress. A *MS* gene from potatoes was cloned and characterized (Zeh et al., 2002). RNA transcription from this *MS* gene was regulated by a day/night rhythm but the protein levels were not altered. Therefore, *MS* was not considered as one of the important components in salt-stress tolerance of plants. However, Narita et al. (2004) reported that HvMS protein levels were increased under salt stress and suggested that *MS* may indeed be important in the salt tolerance of higher plants. Furthermore, they found that this gene complemented a yeast mutant lacking the ability to synthesize methionine under both non-stress and high-salinity conditions.

A *late-embryogenesis-abundant (LEA)* gene family was induced by osmotic condition, dehydration, salt and ABA treatments. The gene family was up-regulated by drought and salinity. Several genes from this gene family were regulated or induced by drought and salinity. These genes include *HVA1* (Hong et al., 1992), *ABA2* and *ABA3* (Gulli et al., 1995), *Paf93* (Grossi et al., 1995), *dehydrins(DHNs)* (Close, 1996) and a *B19* gene family (Hollung et al., 1994). Cattivelli et al. (2002) have summarized this gene family isolated from barley.

Accumulation of glycinebetaine is one mechanism of the barley response to drought or salt stress. Ishitani et al. (1995) reported a gene up-regulated by drought stress and encoding an enzyme of known function – *BADH*. This enzyme is the last step in the betaine synthesis pathway. The mRNA level of *BADH* increased significantly when barley was subjected to drought or saline conditions. Sorbitol plays a role in osmoregulation. The enzyme aldose reductase involved in the accumulation of sorbitol was regulated by ABA (Bartels et al., 1991).

However, the application of the knowledge about the candidate genes for the improvement of salinity tolerance is very limited. One possible reason may be that the biochemical functions of the genes are unknown and the other is that many experiments are not designed to expose the genes that confer salinity tolerance to natural conditions (Munns, 2005).

9. IMPROVEMENT OF SALINITY TOLERANCE BY TRANSGENETIC APPROACHES

There has been substantial progress in identifying genes for tolerance to various abiotic stresses. An alanine aminotransferase isolated from barley roots could increase the drought tolerance of transformed tobacco (Muench and Good, 1994). Xu et al. *(*1996) used a transgenic approach to investigate the function of the HVA1 protein in the stress protection of rice. HVA1 is a group 3 LEA protein that is expressed in the barley aleurone and embryo during late seed development correlating with the seed desiccation stage (Hong et al., 1988). The transgenic rice plants exhibited a high constitutively expression of the HVA1 protein in leaves and roots. The progeny of three transgenic plants was used for the evaluation of the growth performance under water deficit and salt stress treatments. The appearance and development of major damage symptoms such as wilting, dying of old leaves and necrosis of young leaves caused by the stress conditions were delayed in the transgenic plants. The better performance of the transgenic lines under stress conditions was correlated with a higher level of HVA1 protein accumulated in the plants.

Although there are some successful reports for improvement of salinity tolerance through the transgenetic approach, in almost all cases reported so far there are no clear supporting data (Flowers, 2004). As plant salinity tolerance is determined by the coordinated action of many genes, it might not be expected to significantly improve salt tolerance by modifying a single gene. The second wave of transformation attempts to transform plants stress-induced regulatory genes (Bhatnagar-Mathur et al., 2008). Many genes involved in stress response can be simultaneously regulated by a single gene encoding stress-inducible transcription factors (Kasuga et al., 1999). These transcription factors activate cascades of genes that act together in enhancing tolerance towards multiple stresses. For example, the genes involved in the ABA-signalling pathways. Salt hypersensitivity was observed in ABF3-overexpressing and ABF4-overexpressing plants at germination and seedling stages, which indicated that the two genes may participate in the salinity response. Pardo et al. (1998) achieved salinity tolerant transgenic plants by over-expressing phosphatase calcineurin protein involved in the signal transduction under salt stress. Similarly a truncated tobacco mitogene-activated protein kinase (MAPK) triggered an oxidative signal cascade resulting in salinity tolerance.

10. BREEDING FOR SALT-TOLERANT BARLEY

As the physiology and the genetics of salt tolerance are so complicated, is it going to be possible to breed salt-tolerant crops? To date, there have been only limited successes. However, attempts to improve salt tolerance have been made using a variety of approaches involving conventional breeding, wide crossing, the use of physiological traits and more recently, marker-assisted selection and the use of transgenic plants. None of these approaches could be said to offer a universal solution. Conventional breeding programs have rarely delivered enhanced salt tolerance (Flowers and Yeo, 1995) while wide crossing generally reduces yield to unacceptably low levels (Yeo and Flowers, 1989). There has been success using physiological criteria as the basis of selection in rice (Dedolph and Hettel, 1997) and such an approach has been advocated for wheat (Munns et al., 2002). An analysis has shown that whilst it is possible to produce a wide range of transgenic plants where some aspects of a trait relating to salt tolerance were altered, none of them has been tested in the field and few claims for success meet even minimal criteria required to demonstrate enhanced salt tolerance (Flowers, 2004). With the development of molecular marker research, MAS provides a more powerful tool in barley breeding for salt tolerance.

Combining DNA technology and advanced statistical methods (Kearsey, 1998), chromosomal regions that contain the genes that determine quantitative traits can be identified. By crossing parents that differ in one or more aspects of salt tolerance (their phenotypes), and then analyzing the phenotype and the genotype of their offsprings, it has been possible to locate QTLs for salt tolerance. For a plant breeder, such QTLs are particularly attractive, as they can, in principle, be developed to produce markers to aid selection. Such markers can be used in the selection of lines following a crossing program and without the need to determine their phenotype or to take all the lines to seed (Asins, 2002).

In the selection of barley for salt tolerance, Koval et al. (2000) investigated the contribution of the gametophyte in the inheritance of salt tolerance by crossing F_3 and BC_1 hybrids of the tolerant cultivars Rannii 1 and Pirkka with the sensitive cultivar K-30356, and found that the progenies of heterozygous plants grown in saline conditions showed elevated salinity tolerance.

A comparison of the BC_1 hybrid progenies showed that the male and female gametophytes contributed to the inheritance of salt tolerance. Gametic selection is highly efficient during the formation of the female gametophyte and the germination of pollen grains on the stigma.

There is one successful report on the development of salinity tolerance in barley through conventional breeding and selection strategy (Zhao and Liu, 2002). The new barley variety Shannong 50245 was derived from a crossing between the salt-tolerant variety CM72 with a local breeding line 26-2. Shannong 50245 showed similar salinity tolerance as CM72 at germination, seedling and adult plant stages. Under salt concentrations of 0.35%–0.56% in the soil, the new breeding variety showed increased grain yield by 18.3%.

REFERENCES

Abadia, A., et al. (1999). Effects of salinity on the photosynthetic pigment composition of barley (*Hordeum vulgare* L.) grown under a triple-line-source sprinkler system in the field. *J. Plant Physiol, 154*, 392-400.

Ali-Dinar, H. M., et al. (1999). Growth, chlorophyll content, photosynthesis and water relations in guava (*Psidium guajava* L.) under salinity and different nitrogen supply. *Gartenbauwissenschaft, 64*, 54-59.

Al-Karaki, G. N. (2001). Germination, sodium and potassium concentrations of barley seeds as influenced by salinity. *J. Plant Nutr, 24*, 511-522.

Allakhverdiev, S. I., et al. (1999). Genetic engineering of the unsaturation of fatty acids in membrane lipids alters the tolerance of *Synechocystis* to salt stress. *Proc. Natl. Acad. Sci. USA, 96*, 5862-5867.

Almansouri, M., et al. (2001). Effect of salt and osmotic stresses on germination in durum wheat (*Triticum durum* Desf.). *Plant Soil, 231*, 243-254.

Amtmann, A., et al. (2004). Potassium nutrition and salt stress, In Blatt, M. R. (Eds.), *Membrane Transport in Plants*, Blackwell Publishing, Oxford.

Appleby, N., et al. (2008). *Discovery and annotation of barley SNP and SSR markers with AutoSNPdb*, Plant & Animal Genomes XVI Conference, San Diego, CA, PAG-XVI (W52).

Asch, F., et al. (2000). Leaf K^+/Na^+ ratio predicts salinity induced yield loss in irrigated rice. *Euphytica, 113*, 109-118.

Ashraf, M., & Khanum, A. (1997). Relationship between ion accumulation and growth in two spring wheat lines differing in salt tolerance at different growth stages. *J. Agron. Crop Sci, 178*, 39-51.

Ashraf, M., & McNeilly, T. (1988). Variability in salt tolerance of nine spring wheat cultivars. *J.*

Agron. Crop Sci, *160*, 14-21.

Aslam, M., et al. (1993). A rapid screening technique for salt tolerance in rice (*Oryza sativa* L.). *Plant Soil*, *150*, 99-107.

Asins M. J. (2002). Present and future of quantitative trait locus analysis in plant breeding. *Plant Breed*, *121*, 281-291.

Atsunori, F., et al. (1998). Na^+/K^+ antiporter in tonoplast vesicles from rice roots. *Plant Cell Physiol*, *39*, 196-201.

Babourina, O., et al. (2000). Effect of sudden salt stress on ion fluxes in intact wheat suspension cells. *Ann. Bot*, *85*, 759-767.

Babourina, O., et al. (2002). Blue light-induced kinetics of H^+ and Ca^{2+} fluxes in etiolated wild-type and phototropin-mutant *Arabidopsis* seedlings. *Proc. Natl. Acad. Sci. USA*, *99*, 2433-2438.

Barr, A. R., et al. (2000). *Marker assisted selection in theory and practice*, 8th International Barley Genetics Symposium, Univ. South Australia, Waite Campus, Adelaide, 167-178.

Basu, H. S., et al. (1990). Effect of variation in the structure of spermine on the association with DNA and the induction of DNA conformational changes. *Biochem. J*, *269*, 329-334.

Bartels, D., et al. (1991). An ABA and GA modulated gene expressed in the barley embryo encodes an aldose reductase related protein. *EMBO J*, *10*, 1037-1043.

Belkhodja, R., et al. (1994). Chlorophyll fluorescence as a possible tool for salinity tolerance screening in barley (*Hordeum vulgare* L.). *Plant Physiol*, *104*, 667-673.

Belkhodja, R., et al. (1999). Effects of salinity on chlorophyll fluorescence and photosynthesis of barley (*Hordeum vulgare* L.) grown under a triple-line-source sprinkler system in the field. *Photosynthetica*, *36*, 375-387.

Ben-Hayyim, G., et al. (1987). Role of internal potassium in maintaining growth of cultured *Citrus* callus on increasing NaCl and $CaCl_2$ concentrations. *Plant Physiol*, *85*, 434-439.

Bewely, J. D., & Black, M. (1982). *Physiology and biochemistry of seeds in relation to germination*. Vol. 2. Springer-Verlag, Berlin, Germany.

Bhandal, I. S. & Malik, C. P. (1988). Potassium estimation, uptake, and its role in the physiology and metabolism of flowering plants. *Internat. Rev. Cytol*, *110*, 205-254.

Bhatnagar-Mathur, P., et al. (2008). Transgenic approaches for abiotic stress tolerance in plants: retrospect and prospects. *Plant Cell Rep*, *27*, 411-424.

Bhat, P. R., et al. (2007). *High-throughput SNP genotyping and mapping in barley provide ample markers for plant breeders*, Plant & Animal Genomes XV Conference, San Diego, CA.

Binzel, M. L., et al. (1988). Intracellular compartmentation of ions in salt adapted tobacco cells. *Plant Physiol*, *86*, 607-614.

Blumwald, E., et al. (2000). Sodium transport in plant cells. *Biochim. Biophys. Acta*, *1465*, 140-151.

Blumwald, E., & Gelli, A. (1997). Secondary inorganic ion transport at the tonoplast. *Adv. Bot. Res*, *25*, 401-417.

Bohnert, H. J., et al. (1995). Adaptation to environment stresses. *Plant Cell*, *7*, 1099-1111.

Bohnert, H. J. & Jensen, R.G. (1996). Strategies for engineering water-stress tolerance in plants. *Trends Biotechnol*, *14*, 89-97.

Briggs, D. E. (1978). *The origin and classification of barleys*, Barley, Chapman & Hall, London, 76-88.

Brownell, P. F. & Crossland, C. J. (1972). The requirements of sodium as a micronutrient by species having the C_4 dicarboxylic photosynthetic pathway. *Plant Physiol*, *49*, 794-797.

Cattivelli, L., et al. (2002). Genetic bases of barley physiological response to stressful conditions, In Slafer, G. A., et al. (Eds.), *Barley science: Recent advances from molecular biology to agronomy of yield and quality*, Food Products Press, New York, 387-411.

Chartzoulakis, K., & Klapaki, G. (2000). Response of two greenhouse pepper hybrids to NaCl salinity during different growth stages. *Sci. Hort*, *86*, 247-260.

Chauhan, R. P. S., et al. (1980). Free proline accumulation in cereals in relation to salt tolerance. *Plant Soil*, *57*, 167-175.

Cheeseman, J. M. (1982). Pump-leak sodium fluxes in low salt corn roots. *J. Membr. Biol*, *70*, 157-164.

Chen, Z. H., et al. (2005). Screening plants for salt tolerance by measuring K^+ flux: a case study for barley. *Plant Cell Environ*, *28*, 1230-1246.

Chen, Z. H., et al. (2007a). Potassium and sodium relations in salinized barley tissues as a basis of differential salt tolerance, *Funct. Plant Biol*, *34*, 150-162.

Chen, Z. H., et al. (2007b). Root plasma membrane transporters controlling K^+/Na^+ homeostasis in salt-stressed barley. *Plant Physiol*, *145*, 1714-1725.

Chhipa, B. R., & Lal, P. (1985). Effect of soil salinity on yield, yield attributes and nutrient uptake by different varieties of wheat. *Anal. Edaf. Agrobiol*, *11*, 1681-1691.

Chhipa, B. R., & Lal, P. (1995). Na^+/K^+ ratios as the basis of salt tolerance in wheat. *Aust. J. Agric. Res*, *46*, 533-539.

Clarkson, D. T., & Hanson, J. B. (1980). The mineral nutrition of higher plants. *Annu. Rev. Plant Physiol*, *31*, 239-298.

Close, T. J. (1996). Dehydrins: emergence of a biochemical role of a family of plant dehydration proteins. *Physiol. Plant*, *97*, 795-803.

Colmer, T. D., et al. (1995). Differential solute regulation in leaf blades of various ages in salt-sensitive wheat and a salt-tolerant wheat × *Lophopyrum elongatum* (Host) A. Löve amphiploid. *Plant Physiol*, *108*(4), 1715-1724.

Colmer, T. D., et al. (2005). Improving salt tolerance of wheat and barley: future prospects. *Aust. J. Exp. Agric*, *45*, 1425-1443.

Cramer, G. R. (1997). Uptake and role of ions in salt tolerance, In Jaiwal, P. K., et al. (Eds.), *Strategies for improving salt tolerance in higher plants*, Oxford & IBH Publ. Pvt. Ltd., New Delhi, 55-86.

Cramer, G. R., et al. (1994). Salt tolerance is not associated with the sodium accumulation of two maize hybrids. *Aust. J. Plant. Physiol*, *21*, 675-692.

Cramer, G. R., et al. (1989). Na-Ca interactions in barley seedlings: relationship to ion transport and growth. *Plant. Cell Environ*, *12*, 551-558.

Cuartero, J., et al. (1992). Selection of donors for salt tolerance in tomato using physiological traits. *New Phytol*, *121*, 63-69.

Cuin, T. A., et al. (2003). Potassium activities in cell compartments of salt-grown barley leaves. *J. Exp. Bot*, *54*, 657-661

Cuin, T. A., & Shabala, S. (2005). Exogenously supplied compatible solutes rapidly ameliorate NaCl-induced potassium efflux from barley roots. *Plant Cell Physiol*, *46*, 1924-1933.

Cuin, T. A., & Shabala, S. (2007). Amino acids regulate salinity-induced potassium efflux in barley root epidermis. *Planta*, *225*, 753-761.

Daniells, I. G., et al. (2001). Relationship between yield of grain sorghum (*Sorghum bicolor*) and soil salinity under field conditions. *Aust. J. Exp. Agric*, *41*, 211-217.

Davenport, R. J. (1998). Mechanisms of toxic sodium influx in wheat, PhD Thesis, University of Cambridge, UK.

Davenport, R. J., et al. (1997) Sodium-calcium interactions in two wheat species differing in salinity tolerance, *Physiol. Plant*, *99*, 323-327.

Dedolph, C., & Hettel, G. (1997). Rice varieties boost yield and improve saline soils: Partners Making a Difference. *IRRI*, Manila, 37.

Dell'Aquila, A., & Spada, P. (1993). The effect of salinity stress upon protein synthesis of germinating wheat embryos. *Ann. Bot*, *72*, 97-101.

Demidchik, V., et al. (2002). Non-selective cation channels in plants. *Annu. Rev. Plant Physiol. Plant Mol. Biol*, *53*, 67-107.

Demidchik, V., et al. (2003). Free oxygen radicals regulate plasma membrane Ca^{2+}- and K^{+}-permeable channels in plant root cells. *J. Cell. Sci*, *116*, 81-88.

D'Oraci, D., & Bagni, N. (1987). *In vitro* interactions between polyamines and pectic substances. *Biochem. Biophys. Res. Commun*, *148*, 1159-1163.

Dietz, K. J., et al. (2001). Significance of the V-type ATPase for the adaptation to stressful growth conditions and its regulation on the molecular and biochemical level. *J. Exp. Bot*, *52*, 1969-1980.

Donovan, T. J., & Day, A. D. (1969). Some effects of high salinity on germination and emergence of barley (*Hordeum vulgare* L. emend Lam.). *Agron. J*, *61*, 236-238.

Dubcovsky, J., et al. (1996). Mapping of the K^+/Na^+ discrimination locus *Knal* in wheat. *Theor. Appl. Genet*, *92*, 448-454.

Dvořák, J., et al. (1994). Enhancement of the salt tolerance of *Triticum turgidum* L. by the *Knal* locus transferred from the *Triticum aestivum* L. chromosome 4D by homoeologous recombination. *Theor. Appl. Genet*, *87*, 872-877.

Eckermann, C., et al. (2000). Plant methionine synthase: new insights into properties and expression. *Biol. Chem*, *381*, 695-703.

Ellis, R. P., et al. (1997). Mapping physiological traits in barley. *New Phytol*, *137*, 149-157.

Ellis, R. P., et al. (2002). Phenotype/genotype associations for yield and salt tolerance in a barley mapping population. *J. Exp. Bot*, *53*, 1163-1176

Elphick, C. H., et al. (2001). Critical role of divalent cations and Na^+ efflux in *Arabidopsis thaliana* salt tolerance. *Plant Cell Environ. 24*, 733-740.

Epstein, E., et al. (1980). Saline culture of crops: a genetic approach. *Science*, *210*, 399-404.

Ershov, P., et al. (2005). 2005. Activity of ion transporters and salt tolerance in barley. *Russian J. Plant Physiol*, *52*, 765-773.

Essah, P. A., et al. (2003). Sodium influx and accumulation in *Arabidopsis*. *Plant Physiol*, *133*, 307-318.

Fedina, I. S., et al. (2002). Light-dark changes in proline content of barley leaves under salt stress. *Biol. Plant*, *45*, 59-63.

Filho, E. G., & Sodek, L. (1988). Effect of salinity on ribonuclease activity of *Vigna unguiculata* cotyledons during germination. *Plant Physiol*, *132*, 307-311.

Flowers, T. J. (2004). Improving crop salt tolerance. *J. Exp. Bot*, *55*, 307-319.

Flowers, T. J., & Hajibagheri, M. A. (2001). Salinity tolerance in *Hordeum vulgare*: ion concentrations in root cells of cultivars differing in salt tolerance. *Plant Soil*, *231*, 1-9.

Flowers, T. J., & Lauchli, A. (1983). Sodium versus potassium: Substitution and compartmentation, In Pirson, A., & Zimmermann, M. H. (Eds.), *Encyclopedia of Plant Physiology*, new series, *Springer-Verlag*, *Berlin*, *15B*, 651-681.

Flowers, T. J., & Yeo, A. R. (1986). Ion relations of plants under drought and salinity. *Aust. J. Plant Physiol*, *13*, 75-91.

Flowers, T. J., & Yeo, A. R. (1995). Breeding for salinity resistance in crop plants: where next?

Aust. J. Plant Physiol, *22*, 875-884.

Flowers, T. J., et al. (1977). The mechanism of salt tolerance in halophytes. *Annu. Rev. Plant Physiol*, *28*, 89-121.

Foolad, M. R. (1997). Genetic basis of physiological traits related to salt tolerance in tomato, *Lycopersicon esculentum* Mill. *Plant Breed*, *116*, 53-58.

Foolad, M. R. (1999). Comparison of salt tolerance during seed germination and vegetative growth in tomato by QTL mapping. *Genome*, *42*(4), 727-734.

Foolad, M. R. (2004). Recent advances in genetics of salt tolerance in tomato. *Plant Cell Tiss. Org*, *76*, 101-119.

Forster, B. P., et al. (1994). Variation in the leaf sodium content of *Hordeum vulgare* (barley) cultivar Maythorpe and its derived mutant cv. Golden Promise. *Heredity*, *73*, 249-253.

Forster, B. P. (2001). Mutation genetics of salt tolerance in barley: an assessment of Golden Promise and other semi-dwarf mutants. *Euphytica*, *120*, 317-328.

Fortmeier, R., & Schubert, S. (1995). Salt tolerance of maize (*Zea mays* L.): the role of sodium exclusion. *Plant Cell Environ*, *18*, 1041-1047.

Fowler, J. L. (1991). Interaction of salinity and temperature on the germination of crambe. *Agron. J*, *83*, 169-172.

Francois, L. E., et al. (1986). Effect of salinity on grain yield and quality, vegetative growth, and germination of semi-dwarf and durum wheat. *Agron. J*, *78*, 1053-1058.

Fricke, W., et al. (1996). The intercellular distribution of vacuolar solutes in the epidermis and mesophyll of barley leaves changes in response to NaCl. *J. Exp*, *Bot*, *47*, 1413-1426.

Garbarino, J., & Dupont F. M. (1988). NaCl induces a Na^+/H^+ antiport in tonoplast vesicles from barley roots. *Plant Physiol*, *86*, 231-236.

García, A., et al. (1995). The effects of selection for sodium transport and of selection for agronomic characteristics upon salt resistance in rice. *Theor. Appl. Genet*, *90*, 1106-1111.

Garthwaite, A. J., et al. (2005). Salt tolerance in wild *Hordeum* species is associated with restricted entry of Na^+ and Cl^- into the shoots. *J. Exp. Bot*, *56*, 2365-2378.

Gorham, J., et al. (1987). Chromosomal location of a K^+/Na^+ discrimination character in the D genome of wheat. *Theor. Appl. Genet*, *74*, 584-588.

Gorham, J., et al. (1990). Salt tolerance in the *Triticeae*: K/Na discrimination in barley. *J. Exp. Bot*, *41*, 1095-1101.

Gorham, J., et al. (1994). Varietal differences in sodium uptake in barley cultivars exposed to soil salinity or salt spray. *J. Exp. Bot*, *45*, 895-901.

Graner, A., et al. (1991). Construction of an RFLP map of barley. *Theor. Appl. Genet*, *83*, 250-256.

Greenway, H., & Munns, R. (1980). Mechanisms of salt tolerance in non-halophytes. *Annu. Rev. Plant. Physiol. Mol. Biol*, *31*, 149-190.

Greenway, H., & Osmond, C. B. (1972). Salt responses of enzymes from species differing in salt tolerance. *Plant Physiol*, *49*, 256-259.

Gregorio, G. B., et al. (2002). Progress in breeding for salinity tolerance and associated abiotic stresses in rice. *Field Crop Res*, *76*, 91-101.

Grossi, M., et al. (1995). Characterization of two barley genes that respond rapidly to dehydration stress. *Plant Sci*, *105*, 71-80.

Guerrier, G. (1988). Comparative phosphatase activity in four species during germination in NaCl media. *J. Plant Nutr*, *11*, 535-546.

Gulli, M., et al. (1995). Isolation and characterization of absisic acid inducible genes in barley seedlings and their responsiveness to environmental stress. *Plant Physiol*, *14*, 89-96.

Hall, J. L., & Flowers, T. J. (1973). The effect of salt on protein synthesis in the halophyte *Suaeda maritima*. *Planta*, *110*, 361-368.

Hare, P. D., et al. (1998). Dissecting the roles of osmolyte accumulation during stress. *Plant Cell Environ*, *21*, 535-553.

Hasegawa, P. M., et al. (2000). Plant cellular and molecular responses to high salinity. *Annu. Rev. Plant Biol*, *51*, 463- 499.

He, T., & Cramer, G. R. (1993). Salt tolerance of rapid-cycling *Brassica* species in relation to potassium-sodium ratio and selectivity at the whole plant and callus levels. *J. Plant Nutr*, *16*, 1263-1277.

Hong, B., et al. (1988). Cloning and characterization of a cDNA encoding a mRNA rapidly induced by ABA in barley aleurone layers. *Plant Mol. Biol*, *11*, 495-506.

Hong, B., et al. (1992). Developmental and organ-specific expression of an ABA- and stress induced protein in barley. *Plant Mol. Biol*, *18*, 663-674.

Hollung, K., et al. (1994). *AnotherLea B19* gene (*Group1Lea*) from barley containing a single 20 amino acid hydrophilic motif. *Plant Mol. Biol*, *25*, 559-564.

Houshmand, S., et al. (2005). Evaluation of salt-tolerant genotypes of durum wheat derived from *in vitro* and field experiments. *Field Crop Res*, *91*, 345-354.

Hsiao, T. C., & Lauchli, A. (1986). Role of potassium in plant-water relations. *Adv. Plant Nutr*, *2*, 281-312.

Huang, C. X., & van Steveninck, R. F. M., (1990). Salinity induced structural changes in meristematic cells of barley roots. *New Phytol*, *115*, 17-22.

Huang, Y., et al. (2006). Differences in physiological traits among salt-stressed barley genotypes. *Commun. Soil Sci. Plant Anal*, *37*, 557-570.

Ishitani, M., et al. (1995). Expression of the betaine aldehyde dehydrogenase gene in barley in response to osmotic stress and abscisic acid. *Plant Mol. Biol*, *27*, 307-315.

Isla, R., et al. (1997). Field screening of barley cultivars to soil salinity using a sprinkler and a drip irrigation system. *Plant Soil*, *197*, 105-117.

Isla, R., et al. (1998). Validity of various physiological traits as screening criteria for salt tolerance in barley. *Field Crop Res*, *58*, 97-107.

Islam, S., et al. (2007). Salt tolerance in a *Hordeum marinum-Triticum aestivum* amphiploid, and its parents. *J. Exp. Bot*, *58*, 1219-1229.

Jacoby, B., & Hanson, J. B. (1985). Controls on $^{22}Na^{+}$ influx in corn roots. *Plant Physiol*, *77*, 930-934.

James, R. A., et al. (2002). Factors affecting CO_2 assimilation, leaf injury and growth in salt-stressed durum wheat, Funct. *Plant Biol*, 29, 1393-1403.

James, R. A., et al. (2006). Photosynthetic capacity is related to the cellular and subcellular partitioning of Na^{+}, K^{+}, and Cl^{-} in salt-affected barley and durum wheat. *Plant Cell Environ*, *29*, 2185-2197.

Janardhan, K. V., et al. (1979). Effect of various K/Na ratios in saline irrigation water on grain yield and ionic composition of wheat. *Curr. Sci*, *48*, 739-771.

Jaradat, A. A., et al. (2004). Genetic diversity in the Batini barley landrace from Oman: II. Response to salinity stress. *Crop Sci*, *44*, 997-1007.

Jensen, C. R. (1982). Effect of soil water osmotic potential on growth and water relationships in barley during soil water depletion. *Irrig. Sci*, *3*, 111-121.

Jiang, Q., et al. (2006). Gas exchange, chlorophyll fluorescence parameters and carbon isotope discrimination of 14 barley genetic lines in response to salinity. *Field Crop Res*, *96*, 269-278.

Jones, R. G. W., et al. (1979). Ionic and osmotic relations in plant cells, In Laidman, D. L., & Jones, R. G. W. (Eds.), *Recent Advances in the Biochemistry of Cereals*, Academic Press, London, 63-103.

Jones, R. G. W., & Storey, R. (1978). Salt stress and comparative physiology in the Gramineae: II. Glycine betaine and proline accumulation in two salt- and water-stressed barley cultivars. *Aust. J. Plant. Physiol*, *5*, 817-829.

Joshi, S., & Nimbalkar, J. D. (1983). Effect of salt stress on growth and yield in *Cajanus cajan* L. *Plant Soil*, *74*, 291-294.

Kasuga, M., et al. (1999). Improving plant drought, salt, and freezing tolerance by gene transfer of a single stress-inducible transcription factor. *Nat. Biotechnol*, *17*, 287-291.

Kawasaki, S., et al. (2001). Gene expression profiles during the initial phase of salt stress in rice. *Plant Cell*, *13*, 889-905.

Kearsey, M. J. (1998). The principles of QTL analysis (a minimal mathematics approach). *J. Exp. Bot*, *49*, 1619-1623.

Khan, M. A., & Rizvi, Y. (1994). Effect of salinity, temperature and growth regulators on the germination and early seedling growth of *Atriplex griffithii* var. Stocksii. *Can. J. Bot*, *72*, 475-479.

Kingsbury, R. W., & Epstein, E. (1984). Selection for salt-resistant spring wheat. *Crop Sci*, *24*, 310-315.

Kingsbury, R. W., et al. (1984). Physiological responses to salinity in selected lines of wheat. *Plant Physiol*, *74*, 417-423.

Kleinhofs, A., et al. (1993). A molecular, isozyme and morphological map of barley (*Hordeum vulgare*) Genome. *Theor. Appl. Genet*, *86*, 705-712.

Kochian, L. V., & Lucas, W. J. (1988). Potassium transport in roots. *Adv. Bot. Res*, *15*, 93-178.

Koval, V. S. (2000). Male and female gametophyte selection of barley for salt tolerance. *Hereditas*, *132*, 1-5.

Koyama, M. L., et al. (2001). Quantitative trait loci for competent physiological traits determining salt tolerance in rice. *Plant Physiol*, *125*, 406-422.

Kramer, D. (1984). Cytological aspects of salt tolerance in higher plants, In Staples, R.C., & Toenniessen, G. H. (Eds.), *Salinity Tolerance in Plants*, John Wiley and Sons, New York, 3-5.

Krishnamurthy, L., et al. (2007). Screening sorghum genotypes for salinity tolerant biomass production. *Euphytica*, *156*, 15-24.

Kronzucker, H. J., et al. (2006). The cytosolic Na^+:K^+ ratio does not explain salinity-induced growth impairment in barley: a dual-tracer study using $^{42}K^+$ and $^{24}Na^+$. *Plant Cell Environ*, *29*, 2228-2237.

Kumar, D. (1995). Salt tolerance in oilseed brassicas-present status and future prospects. *Plant Breed. Abst*, *65*, 1438-1447.

Lahaye, P. A., & Epstein, E. (1971). Calcium and salt toleration by bean plants. *Physiol. Plant*, *25*, 223-218.

Leigh, R. A. (2001). Potassium homeostasis and membrane transport. *J. Plant Nutri. Soil Sci*, *164*, 193-198.

Leonova, T. G., et al. (2005). Characteristics of salt-tolerant and salt-susceptible cultivars of barley. *Russ. J. Plant Physiol*, *52*, 774-778.

Loreto, F., et al. (2003). Photosynthetic limitations in olive cultivars with different sensitivity to salt stress. *Plant Cell Environ*, *26*, 595-601.

Lin, H. X., et al. (2004). QTLs for Na^+ and K^+ uptake of the shoots and roots controlling rice salt tolerance. *Theor. Appl. Genet*, *108*, 253-260.

Lu, Y. M., et al. (2002). Physiological-biochemistrical and molecular biological analysis of salt-tolerant new lines of *Hordeum brevisubulatium. Sci. Agric. Sin*, *35*, 282-286.

Lucas, W. J., & Kochian, L. V. (1986). Ion transport processes in corn roots: an approach utilizing microelectrode techniques, In Gensler, W. G. (Eds.), *Advanced Agricultural Instrumentation: Design and Use*, Martinus Nijhoff, Dordrecht, The Netherlands, 402-425.

Ludlow, M. M., & Muchow, R. C. (1990). A critical evaluation of traits for improving crop yields in water-limited environments. *Adv. Agron*, *25*, 107-153.

Lynch, J., & Läuchli, A. (1985). Salt stress disturbs the calcium nutrition of barley (*Hordeum vulgare* L.). *New Phytol*, *99*, 345-354.

Lynch, J., & Lauchli, A. (1988). Salinity affects intracellular calcium in corn root protoplasts. *Plant Physiol*, *87*, 351-356.

Lynch, J., et al. (1988). Effects of salinity on the extensibility and Ca availability in the expanding region of growing barley leaves. *Bot. Acta*, *101*, 355-361.

Maas, E. V., & Hoffman, G. J. (1977). Crop salt tolerance-current assessment. *J. Irrig. Drain. Div*, *103*, 115-134.

Maathuis, F. J. M., et al. (1992). Sodium chloride compartmentation in leaf vacuoles of the halophyte *Suaeda maritima* (L.) Dum. and its relation to tonoplast permeability. *J. Exp. Bot*, *43*, 1219-1223.

Maathuis, F. J. M., et al. (1997). Roles of higher plant K^+ channels. *Plant Physiol*, *114*, 1141-1149.

Malkit, A., et al. (2002). Salt induction of fatty acid elongase and membrane lipid modifications in the extreme halo-tolerant alga *Dunaliella salina. Plant Physiol*, *129*, 1320-1329.

Mano, Y., & Takeda, K. (1995). Varietal variation and effects of some major genes on salt tolerance in barley seedlings, Bull. Res. Inst. Bioresour. *Okayama Univ*, *3*, 71-81.

Mano, Y., & Takeda, K. (1997a). Diallel analysis of salt tolerance at germination and the seedling stage in barley (*Hordeum vulgare* L.). *Breed. Sci*, *47*, 203-209.

Mano, Y., & Takeda, K. (1997b). Mapping quantitative trait loci for salt tolerance at germination and the seedling stage in barley (*Hordeum vulgare* L.). *Euphytica*, *94*, 263-272.

Mano, Y., & Takeda, K. (1998). Genetic resources of salt tolerance in wild *Hordeum* species. *Euphytica*, *103*, 137-141.

Marschner, H. (1995). *Mineral nutrition of higher plants*, 2nd edn., Academic Press, London.

Mansour, M. M. F., et al. (1992). Plasma membrane lipid alterations induced by NaCl in winter wheat roots. *Physiol Plant*, *92*, 473-478.

Mäser, P., et al. (2002a). Molecular mechanisms of potassium and sodium uptake in plants. *Plant Soil*, *247*, 43-54.

Maser, P., et al. (2002b). Altered shoot/root Na^+ distribution and bifurcating salt sensitivity in *Arabidopsis* by genetic disruption of the Na^+ transporter *AtHKT1*. *FEBS Lett*, *531*, 157-161.

Maxwell, K., & Johnson, G.N. (2000). Chlorophyll fluorescence: a practical guide. *J. Exp. Bot*, *51*, 659-668.

Mladenova, Y. I. (1990). Influence of salt stress on primary metabolism of *Zea mays* L. seedlings of model genotypes. *Plant Soil*, *123*, 217-222.

Monforte, A. J., et al. (1997). 1997. Salt tolerance in *Lycopersicon* species: 6. Genotype by salinity interaction in quantitative trait loci detection: constitutive and response QTLs. *Theor. Appl. Genet*, *95*, 706-713.

Moons, A., et al. (1995). Molecular and physiological responses to abscisic acid and salts in roots of salt-sensitive and salt-toolerant *indica* rice varieties. *Plant Physiol*, *107*, 177-186.

Morgan, J. M. (1984). Osmoregulation and water stress in higher plants. *Annu. Rev. Plant Physiol*, *25*, 299-319.

Muench, D. G., & Good, A. G. (1994). Hypoxically inducible barley alanine aminotransferase: cDNA cloning and expression analysis. *Plant Mol. Biol*, *24*, 417-427.

Munns, R. (1993). Physiological processes limiting plant growth in saline soils: some dogmas and hypotheses. *Plant Cell Environ*, *16*, 15-24.

Munns, R. (2002). Comparative physiology of salt and water stress. *Plant Cell Environ*, *25*, 239-250.

Munns, R. (2005). Genes and salt tolerance: bringing them together. *New Phytol*, *167*, 645-663.

Munns, R., & James, R. A. (2003). Screening methods for salinity tolerance: a case study with tetraploid wheat. *Plant Soil*, *253*, 201-218.

Munns, R., et al. (1995). The significance of a two-phase growth response to salinity in wheat and barley. *Aust. J. Plant Physiol*, *22*, 561-569.

Munns, R., et al. (2000). Genetic variation for improving the salt tolerance of durum wheat. *Aust. J. Agric. Res*, *51*, 69-74.

Munns, R., et al. (2002). Avenues for increasing salt tolerance of crops, and the role of physiologically based selection traits. *Plant Soil*, *247*, 93-105.

Munns, R., et al. (2006). Approaches to increasing the salt tolerance of wheat and other cereals. *J. Exp. Bot*, *57*, 1025-1043.

Muramoto, Y., et al. (1999). Enhanced expression of a nuclease gene in leaves of barley plants under salt stress. *Gene*, *234*, 315-321.

Muranaka, S., et al. (2002). A salt-tolerant cultivar of wheat maintains photosynthetic activity by suppressing sodium uptake. *Photosynthetica*, *40*, 509-515.

Murillo-Amador, B., et al. (2001). Salt tolerance of cowpea genotypes in the emergence stage.

Aust. J. Exp. Agric, *41*, 81-88.

Nakamura, T., et al. (1996). Differences in mechanisms of salt tolerance between rice and barley plants. *Soil Sci. Plant Nutr*, *42*, 303-314.

Narita, Y., et al. (2004). Characterization of the salt-inducible methionine synthase from barley leaves. *Plant Sci*, *167*, 1009-1016.

Nassery, H. (1979). Salt-induced loss of potassium from plant roots. *New Phytol*, *83*, 27-32

Nevo, E., et al. (1984). Genetic resources of wild cereals in Israel and the vicinity: II. Phenotypic variation within and between populations of wild barley, *Hordeum spontaneum*. *Euphytica*, *33*, 737-756.

Nevo, E. (1992). Origin, evolution, population genetics and resources for breeding of wild barley, In Shewry, P. R. (Eds.), *Hordeum spontaneum* in the fertile crescent, *Barley genetics, biochemistry, molecular biology and biotechnology*, CAB International, Wallingford, UK, 19-43.

Newman, I. A. (2001). Ion transport in roots: measurement of fluxes using ion-selective microelectrodes to characterise transporter function. *Plant Cell Environ*, *24*, 1-14.

Noble, C. L., & Rogers, M. E. (1992). Arguments for the use of physiological criteria for improving the salt tolerance in crops. *Plant Soil*, *146*, 99-107.

Norberg, P., & Liljenberg, C. (1991). Lipids of plasma membranes prepared from oat root cells: effect of induced water deficit tolerance. *Plant Physiol*, *96*, 1136-1141.

Nublat, A., et al. (2001). sas1, an *Arabidopsis* mutant over-accumulating sodium in the shoot, shows deficiency in the control of the root radial transport of sodium. *Plant Cell*, *13*, 125-137.

Ozturk, Z. N., et al. (2002). Monitoring large-scale changes in transcript abundance in drought- and salt-stressed barley. *Plant Mol. Biol*, *48*, 551-573.

Palmgren, M. G. (2001). Plant plasma membrane H^+-ATPases: powerhouses for nutrient uptake. *Annu. Rev. Plant Physiol. Plant Mol. Biol*, *52*, 817-845.

Pang, J. Y., et al. (2006a). Microelectrode ion and O_2 flux measurements reveal differential sensitivity of barley root tissues to hypoxia. *Plant Cell Environ*, *29*, 1107-1121.

Pang, J. Y., et al. (2006b). Effect of secondary metabolites associated with anaerobic soil conditions on ion fluxes and electrophysiology on barley roots. *Plant Physiol*, *145*, 266-276.

Pardo, J. M., et al. (1998). Stress signaling through Ca^{2+}/calmodulin-dependent protein phosphatase calcineurin mediates salt adaptation in plants. *Proc. Natl. Acad. Sci. USA*, *95*, 9681-9686.

Petruzzelli, L., et al. (1992). Physiological and ultrastructural changes in isolated wheat embryos during salt and osmotic shock. *Ann. Bot*, *69*, 25-31.

Poljakoff-Mayber, A., et al. (1994). Seeds of *Kosteletzkya virginica* (Malvaceae): their structure, germination and salt tolerance. *Am. J. Bot*, *81*, 54-59.

Ramagopal, S. (1988). Regulation of protein synthesis in root, shoot and embryonic tissues of germinating barley during salinity stress. *Plant Cell Environ*, *11*, 501-515.

Rana, R. S. (1986). Genetic diversity for salt-stress resistance of wheat in India. *Rachis*, *5*, 32-37.

Rawson, H. M., et al. (1988). An examination of selection criteria for salt tolerance in wheat, barley and triticale genotype. *Aust. J. Agric. Res*, *39*, 759-772.

Rebetzke, G. J., et al. (2000). A hand-held porometer for rapid assessment of leaf conductance in wheat. *Crop Sci*, *40*, 277-280.

Rengel, Z. (1992). The role of calcium in salt toxicity. *Plant Cell Environ*, *15*, 625-632.

Rhode, D., & Hanson, A.D. (1993). Quaternary ammonium and tertiary sulfonium compounds in higher plants. *Annu. Rev. Plant Physiol. Plant Mol. Biol*, *44*, 357-384.

Richards, R. A. (1983). Should selection for yield in saline regions be made on saline or in non-saline soils? *Euphytica*, *32*, 431-438.

Rivelli, A. R., et al. (2002). Effect of salinity on water relations and growth of wheat genotypes with contrasting sodium uptake. *Funct. Plant Biol*, *29*, 1065-1074.

Rostoks, N., et al. (2005). Genome-wide SNP discovery and linkage analysis in barley based on genes responsive to abiotic stress. *Mol. Genet. Genom*, *274(5*), 515-527.

Royo, A., & Aragüés, R. (1993). Validation of salinity crop production functions obtained with the triple line source sprinkler system. *Agron. J*, *85*, 795-800.

Royo, A., & Aragüés, R. (1999). Salinity-yield response functions of barley genotypes assessed with a triple line source sprinkler system. *Plant Soil*, *209*, 9-20.

Santa-Cruz, A., et al. (1999). Short-term salt tolerance mechanisms in differentially salt-tolerant tomato species. *Plant Physiol. Biochem*, *37*, 65-71.

Sayed, H. I. (1985). Diversity of salt tolerance in a germplasm collection of wheat (*Triticum* spp.). *Theor. Appl. Genet*, *69*, 651-657.

Sayed, O. H. (2003). Chlorophyll fluorescence as a tool in cereal crop research. *Photosynthetica*, *41*, 321-330.

Schachtman, D. P., & Liu, W. (1999). Molecular pieces to the puzzle of the interaction between potassium and sodium uptake in plants. *Trends Plant Sci*, *4*, 281-287.

Schubert, S., & Lauchli, A. (1990). Sodium exclusion meehanisms at the root surface of two maize cultivars. *Plant Soil*, *123*, 205-209.

Serafini-Fracassini, D.et al. (1995). Plant transglutaminases. *Phytochemistry*, *40*, 355-365.

Serraj, R., & Sinclair, T. R. (2002). Osmolyte accumulation: can it really help increase crop yield under drought conditions?. *Plant Cell Environ*, *25*, 333-341.

Serrano, R., et al. (1999). A glimpse of the mechanisms of ion homeostasis during salt stress. *J.*

Exp. Bot, *50*, 1023-1036.

Shabala, L., et al. (2005a). Salinity-induced ion flux patterns from the excised roots of *Arabidopsis* sos mutants. *Planta*, *222*, 1041-1050.

Shabala, S. (1996). Leaf temperature kinetics measure plant adaptation to extreme high temperatures. *Aust. J. Plant. Physiol*, *23*, 445-452.

Shabala, S. (1997). H^+ flux kinetics around plant roots after short-term exposure to low temperature: Identifying critical temperatures for plant chilling tolerance. *Plant Cell Environ*, *20*, 1401-1410.

Shabala, S. (2000). Ionic and osmotic components of salt stress specifically modulate net ion fluxes from bean leaf mesophyll. *Plant Cell Environ*, *23*, 825-837.

Shabala, S. (2002). Screening plant for environmental fitness: chlorophyll fluorescence as a "Holy Grail" for plant breeders, In Hemantaranjan, A. (Eds.), *Advances in Plant Physiology*, Scientific Publishers, Jodhpur, India, 287-340.

Shabala, S. (2003). Regulation of potassium transport in leaves: from molecular to tissue level. *Ann. Bot*, *92*, 627-634.

Shabala, S. (2006). Non-invasive microelectrode ion flux measurements in plant stress physiology, In Volkov, A. (Eds.), *Plant electrophysiology - theory and methods*, Springer-Verlag, Berlin, Germany, 35-71.

Shabala, S., & Hariadi, Y. (2005). Effects of magnesium availability on the activity of plasma membrane ion transporters and light-induced responses from broad bean leaf mesophyll. *Planta*, *221*, 56-65.

Shabala, S. (2002). Knowles, A., Rhythmic patterns of nutrient acquisition by wheat roots. *Funct. Plant Biol*, *29*, 595-605.

Shabala, S., & Lew, R. R. (2002). Turgor regulation in osmotically stressed *Arabidopsis thaliana* epidermal root cells: direct support for the role of inorganic ion uptake as revealed by concurrent flux and cell turgor measurements. *Plant Physiol*, *129*, 290-299.

Shabala, S., & Newman, I. A. (2000). Salinity effects on the activity of plasma membrane H^+ and Ca^{2+} transporters in bean leaf mesophyll: masking role of the cell wall. *Ann. Bot*, *85*, 681-686.

Shabala, S., et al. (1998). Salinity effect on bioelectric activity, growth, Na^+ accumulation and chlorophyll fluorescence of maize leaves: a comparative survey and prospects for screening. *Aust. J. Plant. Physiol*, *25*, 609-616.

Shabala, S., et al. (2000). Ion-specific mechanism of osmo-regulation in bean mesophyll cells. *J. Exp. Bot*, *51*, 1243-1253.

Shabala, S., et al. (2003). Effect of calcium on root development and root ion fluxes in salinized barley seedlings. *Funct. Plant Biol*, *30*, 507-514.

Shabala, S., et al. (2005). Effect of divalent cations on ion fluxes and leaf photochemistry in salinized barley leaves. *J. Exp. Bot*, *56*, 1369-1378.

Shabala, S., et al. (2006a). Extracellular Ca^{2+} ameliorates NaCl-induced K^+ loss from *Arabidopsis* root and leaf cells by controlling plasma membrane K^+-permeable channels. *Plant Physiol*, *141*, 1653-1665.

Shabala, S., et al. (2006b). Oscillations in plant membrane transport: Model predictions, experimental validation, and physiological implications. *J. Exp. Bot*, *57*, 171-184.

Shannon, M. C. (1997). Adaptation of plants to salinity. *Adv. Agron*, *60*, 75-120.

Shannon, M. C., & Noble, C. L. (1990). Genetic approaches for developing economic salt-tolerant crops, *Agricultural salinity assessment and management*, ACSE Manuals and reports on engineering practice No. 71, (Tanji, K. K., ed.). ASCE, New York., 161-185.

Shi, W. M., et al. (2001). Cloning of peroxisomal ascorbate peroxidase gene from barley and enhanced thermotolerance by overexpressing in *Arabidopsis thaliana*. *Gene*, *273*, 23-27.

Somerville, C. (1995). Direct tests of the role of membrane lipid composition in low-temperature-induced photo-inhibition and chilling sensitivity in plants and cyanobacteria. *Proc. Natl. Acad. Sci. USA*, *92*, 6215-6218.

Srivastava, J. P., & Jana, S. (1984). Screening wheat and barley germplasm for salt tolerance, *Salinity tolerance in plants: strategies for crop improvement*, In Staples, R. C., & Toenniessen, G.H. (Eds.), Wiley, New York, USA, 273-283.

Storey, R., & Jones, R. G. (1978). Salt stress and comparative physiology in the Gramineae, I. Ion relations of two salt- and water-stressed barley cultivars, California Mariout and Arimar. *Aust. J. Plant Physiol*, *5*, 801-816.

Subbarao, G. V., et al. (2003). Sodium: a functional plant nutrient. *Crit. Rev. Plant Sci*, *22*, 391-416.

Suhayda, G. G., et al. (1992). Comparative response of cultivated and wild barley species to salinity stress and calcium supply. *Crop Sci*, *32*, 154-163.

Tajbakhsh, M., et al. (2006). Physiological and cytological response of salt-tolerant and non-tolerant barley to salinity during germination and early growth. *Aust. J. Exp. Agric*, *46*, 555-562.

Takuji, N., et al. (1996). Differences in mechanisms of salt tolerance between rice and barley plants. *Soil Sci. Plant Nutri*, *42*, 303-314.

Tal, M. (1985). Genetics of salt tolerance in higher plants: theoretical and practical considerations. *Plant Soil*, *89*, 199-226.

Tassoni, A., et al. (1996). Polyamine binding to plasma membrane vesicles from zucchini hypocotyls. *Plant Physiol*, *110*, 817-824.

Tester, M., & Davenport, R. (2003). Na^+ tolerance and Na^+ transport in higher plants. *Ann. Bot*, *91*, 503-527.

Thalji, T., & Shalaldeh, G. (2007). Screening wheat and barley genotypes for salinity resistance. *J. Agron*, *6*, 75-80.

Tobe, K., et al. (2003). Alleviatory effects of calcium on the toxicity of sodium, potassium and magnesium chlorides to seed germination in three non-halophytes. *Seed Sci. Res*, *13*, 47-54.

Tyerman, S. D., et al. (2001). Oscillations in proton transport revealed from simultaneous measurements of net current and net proton fluxes from isolated root protoplasts: MIFE meets patch-clamp. *Aust. J. Plant Physiol*, *28*, 591-604.

Ueda, A., et al. (2004). Osmotic stress in barley regulates expression of a different set of genes than salt stress does. *J. Exp. Bot*, *55*, 2213-2218.

Uozumi, N., et al. (2000). The *Arabidopsis HKT1* gene homolog mediates inward Na^+ currents in *Xenopus laevis* oocytes and Na^+ uptake in *Saccharomyces cerevisiae. Plant Physiol*, *122*, 1249-1260.

Varshney, R. K., et al. (2004). Molecular marker maps of barley: a resource for intra- and interspecific genomics, In Lorz, H., & Wenzel, G. (Eds.), *Biotechnology in agriculture and forestry 55: Molecular markers systems in Plant Breeding and Crop Improvement*, Springer, Heidelberg, Germany, 229-243.

Verslues, P. E., et al. (2006). Methods and concepts in quantifying resistance to drought, salt and freezing, abiotic stresses that affect plant water status. *Plant J*, *45*, 523-539.

Véry, A. A., & Sentenac, H. (2002). Cation channels in the *Arabidopsis* plasma membrane. *Trends Plant Sci*, *7*, 168-175.

Volkov, V., et al. (2004). *Thellungiella halophila*, a salt-tolerant relative of *Arabidopsis thaliana*, possesses effective mechanisms to discriminate between potassium and sodium. *Plant Cell Environ*, *27*, 1-14.

Wagner, G. J. (1982). Compartmentation in plant cells: The role of the vacuole, In Creasy, L. L., & Hrazdina, G. (Eds.), *Cellular and Subcellular Localisation in Plant Metabolism*, Plenum Press, New York., 1-45.

Walia, H., et al. (2006). Expression analysis of barley (*Hordeum vulgare* L.) during salinity stress. *Funct. Integr. Genom*, *6*, 143-156.

Wang, Y., & Nil, N. (2000). Changes in chlorophyll, ribulose biphosphate carboxylase-oxygenase, glycine betaine content, photosynthesis and transpiration in *Amaranthus tricolor* leaves during salt stress. *J. Hortic. Sci. Biotechnol*, *75*, 623-627.

Wei, W. X., et al. (2001). Variation between two near isogenic barley (*Hordeum vulgare*) cultivars in expression of the B subunit of the vacuolar ATPase in response to salinity. *Hereditas*, *135*,

227-231.

Wei, W. X., et al. (2003). Salinity induced differences in growth, ion distribution and partitioning in barley between the cultivar Maythorpe and its derived mutant Golden Promise. *Plant Soil*, *250*, 183-191.

Weidner, A., et al. (2007). *QTLs for salt resistance vary with development stage in field-grown barley*, *International Research on Food Security*, Natural Resource Management and Rural Development, Proc. Tropentag 2007, Cuvillier Verlag, Gottingen.

Wenzl, P., et al. (2006). A high-density consensus map of barley linking DArT markers to SSR, RFLP and STS loci and phenotypic traits. *BMC Genom*, *7*, 206.

Wherrett, T., et al. (2005). Effect of aluminium on membrane potential and ion fluxes at the apices of wheat roots. *Funct. Plant Biol*, *32*, 199-208.

White, P. J., & Davenport, R. J. (2002). The voltage independent cation channel in the plasma membrane of wheat roots is permeable to divalent cations and may be involved in cytosolic Ca^{2+} homeostasis. *Plant Physiol*, *130*, 1386-1395.

Wolf, O., et al. (1990). Concentrations and transport of solutes in xylem and phloem along the leaf axis of NaCl-treated *Hordeum vulgare*. *J. Exp. Bot*, *41*, 1133-1141.

Xiong, L., & Zhu, J. K. (2002). Molecular and genetic aspects of plant responses to osmotic stress. *Plant Cell Environ*, *25*, 131-139.

Xu, D., et al. (1996). Expression of a late embryogenesis abundant protein gene, *HVAJ*, from barley confers tolerance to water deficit and salt stress in transgenic rice. *Plant Physiol*, *110*, 249-257.

Yamaguchi, M., & Kasamo, K. (2001). Modulation in the activity of purified tonoplast H+-ATPase by tonoplast glycolipids prepared from cultured rice (*Oryza sativa* L. var. Boro) cells, *Plant Cell Physiol*, *42*, 516-523.

Yamaguchi, T., & Blumwald, E. (2005). Developing salt-tolerant crop plants: challenges and opportunities. *Trends Plant Sci*, *10*(12), 615-620.

Yamamoto, A., et al. (2004). Effect of difference in nitrogen media on salt-stress response and contents of nitrogen compounds in rice seedlings. *Soil Sci. Plant Nutr*, *50*, 85-93.

Yancey, P. H., et al. (1982). Living with water stress: evolution of osmolyte systems. *Science*, *217*, 1214-1222.

Yapsanis, T., et al. (1994). Protein phosphorylation dephosphorylation in alfalfa seeds germinating under salt stress. *J. Plant Physiol*, *143*, 234-240.

Ye, J. M., et al. (1987). Screening salt-tolerant barley genotypes via F_1 anther culture in salt stress media. *Theor. Appl. Genet*, *74*, 426-429.

Yeo, A. R., & Flowers, T. J. (1986). Salinity resistance in rice (*Oryza sativa* L.) and a pyramiding approach to breeding varieties for saline soils. *Aust. J. Plant Physiol*, *13*, 161-173.

Yeo, A. R., & Flowers T. J. (1989). Selection for physiological characters - examples from breeding for salt tolerance, In Jones, H. G., Flowers, T. J., & Jones, M. B. (Eds.), *Plants under Stress Biochemistry*, Physiology and Ecology and their Application to Plant Improvement, Cambridge University Press, Cambridge, 217-234.

Zeh, M., et al. (2002). Cloning and characterization of a cDNA encoding a cobalamin-independent methionine synthase from potato (*Solanum tuberosum* L.). *Plant Mol. Biol*, *48*, 255-265.

Zeng, L., et al. (2002). Evaluation of salt tolerance in rice genotypes by multiple agronomic parameters. *Euphytica*, *127*, 235-245.

Zhao, F. G., & Qin, P. (2005). Protective effects of exogenous fatty acids on root tonoplast function against salt stress in barley seedlings. *Environ. Exp. Bot*, *53*, 215-223.

Zhao, F. G., et al. (2000). Effects of salinity stress on the levels of covalently and non-covalently bound polyamines in plasma membrane and tonoplast isolated from leaves and roots of barley seedlings. *Acta. Bot. Sin*, *42*, 920-926.

Zhao, T. F., & Liu, A. F. (2002). A new salt-tolerant barley line Shannong 50245. *China Seed Industry*, *7*, 28.

Zhong, G. Y., et al. (1995). Evidence for common genetic mechanisms controlling the tolerance of sudden salt stress in the tribe Triticeae. *Plant Breed*, *114*, 297-302.

Zhu, G. Y., et al. (2001). Characterisation of rice (*Oryza sativa* L.) F_3 populations selected for salt resistance. I. Physiological behaviour during vegetative growth. *Euphytica*, *121*, 251-263.

Zhu, J. K. (2000). Genetic analysis of plant salt tolerance using *Arabidopsis*. *Plant Physiol*, *124*, 941-948.

Zhu, J. K. (2001). Plant salt tolerance. *Trends Plant Sci*, *6*, 66-71.

Zhu, J. K., et al. (1998). Genetic analysis of salt tolerance in *Arabidopsis*: evidence for a critical role of potassium nutrition. *Plant Cell*, *10*, 1181-1191.

Zidan, I., et al. (1991). Sodium does not compete with calcium in saturating plasma membrane sites regulating ^{22}Na influx into salinized maize roots. *Plant Physiol*, *96*, 331-334.

Zid, E., & Grignon, C. (1985). Sodium-calcium interactions in leaves of *Citrus aurantium* grown in the presence of NaCl. *Physiol. Veg*, *23*, 895-903.

Živanović, B. D., et al. (2005). Light-induced transient ion flux responses from maize leaves and their association with leaf growth and photosynthesis. *Plant Cell Environ*, *28*, 340-352.

CHAPTER 7.

GENETICS AND BREEDING SOYBEAN FOR SALT TOLERANCE

Rongxia Guan, Lei Tian, Xikui Zhang, Ruzhen Chang and Lijuan Qiu*

National Key Facility for Crop Gene Resources and Genetic Improvement, Key Laboratory of Crop Germplasm & Biotechnology, MOA/ Institute of Crop Sciences, Chinese Academy of Agricultural Sciences, Beijing 100081, China

*Correspondence: qiu_lijuan@263.net

1. INTRODUCTION

Soybean (*Glycine max* L.) is one of the most important oil-seed crops in the world. The demand for soybean is continuously rising due to the ever-growing world population. However, extensive tracts of formerly arable land are being lost from crop production due to increased soil salinity in arid and semi-arid areas. It is estimated that 1,000 Mha of the earth's surface are covered by saline soils. A soil containing excess soluble salts to impair plant growth, i.e. with an EC greater than 40 mmol L^{-1} NaCl (about 4 dS m^{-1}) is considered saline (Stoddard et al., 2006). According to projections of Blumwald and Grover (2006) about 50% of the arable land will be affected by salt stress by the year 2050. In China, 8 Mha of agricultural land are becoming increasingly salinized due to improper management of irrigation and fertilization (Gao and Lin, 2005). This chapter describes briefly the salinity effects on soybean plants, physiological mechanisms of salt tolerance in soybean, identification of salt-tolerant soybean germplasm, inheritance of salt tolerance, mapping and cloning of candidate genes for salt tolerance and perspectives of breeding soybean for salt tolerance.

2. SALINITY EFFECTS ON SOYBEAN

2.1 Plant growth and yield

Soybean is classified as a salt-sensitive glycophyte (Lauchli, 1984). The salinity threshold for soybean is set at $EC_e = 5.0$ dS m^{-1}. In soybean, salinity stress inhibits seed germination and seedling growth. A significant increase in mortality rate and leaf necrosis were induced when EC_e increased to 10.2 dS m^{-1} (Abel and MacKeinze, 1964). Soybean relative yield is dramatically decreased from 100% at 0.8 dS m^{-1} to 80% at 4.0 dS m^{-1} and 44% at 6.7 dS m^{-1}. Soil salinity caused a reduction of the relative biological nitrogen contribution from 100% at 0.8 dS m^{-1} to 77% at 4.0 dS m^{-1} and 28% at 6.7 dS m^{-1} (Katerji et al., 2003).

2.2 Seed quality

Seed protein content decreased and oil content increased significantly under relatively low salt stress (14–15 dS m^{-1}). Linoleic and linolenic acids content increased and oleic acid decreased significantly. However, protein content increased and oil content decreased under high salt stress (18–20 dS m^{-1}). High linolenic acid and protein contents may increase the salt tolerance of soybean (Chang et al., 1994).

2.3 Nodulation

Soybean nodulation is also adversely affected by salt stress. Studies have shown that salinity significantly decreases nodule number and dry-weight (Bernstein and Ogata, 1966). The availability of oxygen to nodules is reduced and the fermentative pathways are stimulated (Serraj et al., 1994). The ability of tissues to supply water to root cells under salt stress is reduced (Joly, 1989). In other studies a nuclear deformation of the meristematic root cells occurred and it was followed by the degradation of nuclei in the apical region of the root tip (Liu et al., 2000).

2.4 Other effects

Salinity stress induced a significant increase of sodium and chloride and reduced the accumulation of potassium, calcium and magnesium in the soybean leaves (Abel, 1969; Essa, 2002).

3. PHYSIOLOGICAL MECHANISMS OF SALT TOLERANCE

The physiological mechanisms of salt tolerance are complex. It is postulated that they consist of two main components (Greenway and Munns, 1980). One component is an ‘osmotic effect', which limits water absorption due to the soil salinity in the rhizosphere. The other component is an 'ionic effect', due to the toxicity from specific excess ions. Salt tolerance in soybean is thought to be primarily related to the ionic component, in which tolerant plants limit the accumulation of excess ions to reduce injury (Lauchli and Wieneke, 1979; Umezawa et al., 2000; 2002). Salt-tolerant soybean genotypes are able: i) to prevent salt ions from moving from the roots to other plant parts; ii) not to accumulate too many salt ions in the leaves and stems; and iii) to have better osmotic adjustment in the plant cells.

Abel (1969) reported that the chloride concentration in the leaves of salt-tolerant soybeans was 10 times less than that of salt-sensitive soybeans. The transport of Cl^- in salt-tolerant soybean cultivars from the root to stems and leaves was exceedingly low (Abel and MacKeinze, 1964). The plants of the salt-tolerant cultivar Lee were taller and maintained lower Na^+ and Cl^- concentrations, a higher K^+ concentration and a higher K^+/Na^+ ratio at higher salinity levels than those of the salt-sensitive cultivars Colquitt and Clark 63 (Essa, 2002). Another salt-tolerant cultivar Dare showed a higher relative shoot and root growth, greater water extraction ability, root pressure, better root osmotic adjustment and less sodium accumulation in the plant tissues than those of the salt-sensitive cultivars. The salt tolerance of Dare was associated to high water uptake, and Na^+ and Cl^- exclusion from being transported from roots to upper portions of plants. Salt tolerance in soybean was also shown to be related to the maintenance of a more stable water content in shoots as well as the accumulation of soluble saccharides and proteins, the amino acid proline, and K^+ and Ca^+ for osmotic adjustment (El-Samad and Shaddad, 1997).

Differences in salt tolerance among *Glycine* species have been reported (Pantalone et al., 1997; Luo et al., 2005a; Kao et al., 2006). Greater variation in NaCl tolerance was found among the perennial *Glycine* accessions than among the *G. max* cultivars. The NaCl tolerance thresholds based on the Weibull model of leaf chlorosis ranged from 3.0 g L^{-1} to 17.5 g L^{-1} NaCl for the perennial accessions whereas only from 5.2 g L^{-1} to 8.0 g L^{-1} NaCl in the *G. max* cultivars (Pantalone et al., 1997). Luo et al. (2005a) reported that the salt tolerance mechanism of the cultivated soybean (*G. max*) was different from that of the wild soybean (*G. soja*). Comparing the differential sensitivity of *G. max* and *G. soja* to Cl^- and Na^+, leaves of *G. max* genotypes were more heavily damaged in the Cl^- solution than in the Na^+ solution. The salt tolerance in *G. max* was mainly due to prevention of Cl^- transport from the roots to the upper portion of the plant, then preventing its toxic accumulation in stems and leaves. In contrast leaves of the salt-tolerant wild soybeans, *G. soja,* were not as susceptible as *G. max* to Cl^- toxicity as that of Na^+. The salt tolerance in *G. soja* resulted primarily from the exclusion of Na^+ from the roots and preventing sodium accumulation at toxic concentrations in the stems and leaves. The descendants of a cross between tolerant wild and cultivated soybean strains were more tolerant to salt stress of NaCl and

Cl^- salts than those of a cross between tolerant *G. max* cultivars (Luo et al., 2005a). This indicates that inter-specific crosses between *G. max* and *G. soja* offer the possibility of improving salt tolerance in soybean cultivars.

4. IDENTIFICATION OF SALT-TOLERANT SOYBEAN GERMPLASM

4.1 Identification and classification methods

Two screening techniques have been used for the selection of salt-tolerant soybeans: i) soil with a high salt content; and ii) hydroponics with high salt addition to a nutrient solution. The field evaluation of genotypes in highly saline soils may be difficult due to the great variability of the salinity across the experimental conditions. The main classification methods for evaluating salt tolerance are the direct phenotype grading and the salt toxicity indexing method (Shao et al., 1986; Ma and Wang, 1994; Lee et al., 2004) as shown in Table 7.1. In these experiments the salt toxic index is:

$$\text{Salt toxic index} = \frac{\sum(\text{individual number with the same grade}\times\text{grade number})\times 100}{\text{surveyed plant number before treatment}\times 5} \tag{7.1}$$

4.2 Screening germplasm for salt tolerance

Shao et al. (1995) evaluated 1,716 Chinese soybean and 415 accessions for salt tolerance with three replications in a field of Shandong province, China. The soil was a sandy loam with moderate saline content. Soybean lines were irrigated with 15–17 dS m^{-1} saline water made from a mixture of fresh water and underground salt water and were screened at the V2 or V3 seedling stage. A total of 85 salt-tolerant Chinese soybean germplasm at the seedling stage were identified whereas only 10 germplasm from the United States were salt-tolerant. Parker et al. (1986) planted 15 soybean genotypes in two fields with a history of leaf scorch symptoms from Cl^- toxicity due to KCl fertilizer applications. The soil types were Leefield sand (arenic Plinthaquic Paleudults) and Alapaha sand (arenic Plinthic Paleaquults). They found that 10 out of 15 soybean cultivars were tolerant to high chloride by leaf scorch rating and leaf Cl^- concentration. Yang and Blanchar (1993) evaluated soybean in the field with a Mexico silt loam soil, with and without added Cl^-. A total of 673 kg ha^{-1} Cl^- was added as a $CaCl_2$ solution to the Cl^- plot. Nineteen out of 60 soybean cultivars were salt-tolerant based on leaf scorch ratings and leaf Cl^- concentration. These screening studies showed that different saline soil conditions can be used for phenotyping soybean genotypes for salt tolerance (Table 7.1).

Hydroponics based on the nutrient solution of Hoagland and Arnon (1938) modified by Johnson et al. (1957) with added NaCl is a widely used method to screen soybeans for salt tolerance. Many genotypes can be screened in a limited space like the greenhouse and the salt level can be easily controlled.

Table 7.1 Summary of soybean screening methods for salt tolerance

Screening techniques	Classification method	Reference
1) Germinating soybean in sand and placing 5 seedlings per replication in nutrient solution 14–21 days after emergence 2) adding NaCl to the solution for 14–31 days	0 = healthy (no apparent symptoms of scorch) 1 = slight scorch (25% of the leaf area showed scorch symptoms) 2 = moderate scorch (50% of the leaf area showed scorch symptoms) 3 = severe (75% of the leaf area showed scorch symptoms) 4 = dead (plants were brown and withered)	Pantalone et al., 1997
1) Plants were grown in 27 cm diameter pots filled with builder's washed sand 2) 14 days after emergence, 100 mmol L^{-1} NaCl was added to the nutrient solution, and this modified solution was applied daily 3) visual ratings of leaf scorching and chlorosis were taken 45 days after emergence	0 = all plants dead 3 = plants with light green or chlorotic appearance 5 = plants with normal green leaves	Lee et al., 2004
1) Soybean was planted in sandy loam soil with moderate saline content 2) at germination and the V2 or V3 seedling stage using 15–17 ds m^{-1} saline water made from a mixture of fresh water and underground salt water 3) plants were screened 3–5 days after treatment	1 = normal 2 = Less than 30% plants showed scorch symptoms 3 = most of the plants showed scorch symptoms, less than half of the plants dead 4 = plant growth was decreased, less than 80% of the plants dead 5 = more than 80% plants dead Grades and equivalent toxic indexes were: 1 = 0%–20% 2 = 20.1%–35% 3 = 35.1%–65% 4 = 65.1%–90% 5 = 90.1%–100%	Shao, 1986

(continued)

Screening techniques	Classification method	Reference
1) Soybean was planted in alkaline chernozem soil with 0.2% saline content 2) at the V3 seedling stage using 10 dS m^{-1} saline water made from a mixture of fresh water, refined salt and alkali 3) plants were screened 5 days after treatment	0 = normal 1 = basically normal, less than 10% leaf area showed scorch symptoms or more than 4 green leaves per plant 2 = basically normal, More than 10% leaf area showed scorch symptoms or more than 2 green leaves 3 = plant growth was decreased, Half of the leaf area showed scorch symptoms or 2 green leaves per plant 4 = plant growth was decreased badly, 75% of leaf area showed scorch symptoms or only one green leaf per plant 5 = nearly dead, more than 75% of leaf area showed scorch symptoms or only one central leaf was alive per plant Grades and equivalent toxic indexes were: 0 = 0 1 = 0.1%–20% 2 = 20.1%–35% 3 = 35.1%–65% 4 = 65.1%–90% 5 = 90.1%–100%	Ma and Wang, 1994

The identification results of salt-tolerant soybean germplasm depend on the genotype, the salt concentration, and other environmental factors such as temperature and light (Table 7.2). Salt concentration is the most critical factor for phenotype grading for salt tolerance. Reported threshold salinity values to detect salt tolerance in soybean are variable among different studies. Chinnusamy et al. (2005) determined in their study that the threshold salinity to detect tolerance was 3.2 dS m^{-1}. The cultivar 'Lee' (salt-tolerant) produced more than twice relative shoot fresh weight and it was significantly lower in chlorosis score than the other genotypes at 6.0 dS m^{-1} salinity level. It was possible to differentiate soybean genotypes for tolerance or sensitivity at 7.5 dS m^{-1} but all genotypes were sensitive at 10.9 dS m^{-1} and had similar leaf chloride levels. The most significant phenotypic differences for salt tolerance between 'Dare' (tolerant) and 'Tachiyutaka' (sensitive) were obtained at 40 mmol L^{-1} NaCl (An et al., 2002). Seventeen soybean cultivars were evaluated for salt tolerance at five locations over two years. Three cultivars with higher yield in the saline fields were selected. It is expected that the area planted with soybean will increase by using the selected cultivars in the saline fields of Hebei province in China (Pan, 2007).

Table 7.2 Salt-tolerant soybean germplasm identified by different investigators

Serial number	No. Germplasm	Tolerant germplasm		Growth stage	Reference
		Number	Ratio (%)		
1	1716	242	14.1	Germination	Shao et al., 1986

(continued)

Serial number	No. Germplasm	Tolerant germplasm		Growth stage	Reference
		Number	Ratio (%)		
2	1716	85	5.0	Seedling	Shao et al., 1986
3	1716	82	4.8	Mature	Shao et al., 1986
4	1514	201	13.3	Seedling	Shao, 1988
5	17	3	17.6	Full-life	Pan, 2007
6	15	10	66.7	Seedling	Parker et al.,1986
7	60	19	31.7	Seedling	Yang and Blanchar, 1993

5. INHERITANCE OF SALT TOLERANCE AND MAPPING GENES FOR SALT TOLERANCE

5.1 Inheritance

To characterize the inheritance of salt tolerance, soybean genotypes were screened for Cl^- inclusion (sensitive) and exclusion (tolerant). A salt solution was added to screen F_2 populations from parents differing in chloride accumulation. The populations were segregated in ratios of three non-necrotic plants very low in chloride to one necrotic plant very high in chloride with a 1:1 segregation ratio from a test cross. It was concluded that the factor to exclude or include chlorides in the soybean leaves and stems is controlled by a single dominant gene. Gene symbols *Ncl* and *ncl* were proposed as the dominant chloride excluder allele for tolerance and the recessive chloride includer for sensitivity, respectively (Abel, 1969). Shao et al. (1994) studied the inheritance of the salt tolerance in soybean based on the segregation performance of the progenies resulting from crossing salt-tolerant and -sensitive cultivars. The F_1 plants expressed salt tolerance and the F_2 populations segregated in a ratio of three tolerant to one sensitive. The results indicated that the salt tolerance was based on a single dominant gene and it was suggested that this gene was the same as that reported by Abel (1969).

5.2 Tagging salt tolerance genes/QTLs

Once the sources of salt tolerance are identified and their genetics are determined, it is important to identify closely linked genetic markers for use in MAS. Zhong et al. (1997) suggested that three DNA amplified fingerprinting (DAF) markers present only in cultivars Morgan and Wenfeng 7 might be linked to salt tolerance. Six RAPD markers were identified only in salt-tolerant soybeans indicating a possible relationship between the molecular markers and salt tolerance

(Zhang et al., 1999). Guo et al. (2000) developed a co-dominant RAPD marker (a 600 bp band in the sensitive parent and a 700 bp band in the tolerant parent) closely linked to a salt tolerance gene but the location of the marker was not determined. Now this RAPD marker has been converted to SCAR marker and mapped to the soybean linkage group N near the SSR marker Satt255 (unpublished data). Lee et al. (2004) mapped a major QTL for salt tolerance in the cultivar S-100 near the marker Sat_091 on linkage group N. This salt tolerance gene accounted for 79% of the genetic variation in combined environments. The association between salt tolerance and the same marker allele was consistent in the related salt-tolerant cultivars. They predicted that the major QTL on LG-N is likely to be the *Ncl* locus initially named by Abel (1969). It, thus, appears that the same gene is relatively common although tests of allelism are yet to be conducted.

6. CLONING OF CANDIDATE GENES FOR SALT TOLERANCE

Molecular analysis has led to the identification of a number of genes induced by NaCl in soybean (Table 7.3). Most of the genes in the functional groups have been identified as salt-inducible under stress conditions. Some of the genes could be induced by other abiotic stress, such as cold and drought (Luo et al., 2005b; Sun et al., 2006). By using cDNA-AFLP, Umezawa et al. (2002) found several *Glycine max stress responsive* (*GSR*) genes such as *GSR-8*, *98*, *110* and *112*, which were induced by the NaCl treatment and showed tissue-specific expression. These authors also suggested that the salt tolerance of soybean is resulting from the response to both ionic and osmotic effects. The gene expression was abundant in the soybean under salt stress. Transcripts, which could be determined from the ionic (NaCl-specific) and osmotic effects (common from NaCl with PEG) were 44% and 40%, respectively. The gene expression dependent from the ionic effect was more abundant in the roots indicating a greater response to the ionic stress in the roots than in the shoots. On the other hand, the *GSR* gene expression from osmotic effects was more abundant in the shoots than in the roots.

Table 7.3 Genes/protein induced by salt stress in soybean

Salt responsive genes/proteins	Characteristic feature	Reference
GmCAX1	Expressed in all tissue of the plants, over expression in *Arabidopsis* accumulated less Na^+, K^+ and Li^+, may function as an antiporter for Na^+, K^+ and Li^+	Luo et al., 2005b
GmDREBa *GmDREBb* *GmDREBc*	*GmDRBa* and *GmDRBb* in soybean leaves were induced by salt, drought, and cold stress, expression of *GmDRBc* was induced in roots by salt, drought, and abscisic acid treatments	Li et al., 2005

(continued)

Salt responsive genes/proteins	Characteristic feature	Reference
GmDREB2	Classified into A-5 subgroup in DREB subfamily in AP2/EREBP family, over expression in *Arabidopsis*, resulting in enhanced tolerance to drought and high-salt stresses	Chen et al., 2007
GmTDFs	A novel cytosolic leucine-zipper-like protein, induced in the stem and lower-expanded leaf	Aoki et al., 2005
GmPAP3	Plays a role in the adaptation of soybean to NaCl stress	Liao et al., 2003
GmNHX1	Expression was increased by ABA treatment, NaCl, KCl, LiCl and dehydration stress, over expression in *Lotus corniculatus* L. got lower Na^+ and K^+ content, and higher K^+/ Na^+ ratio than that in the control line	Sun et al., 2006
STL	Shared 63.6% identity with *Arabidopsis thaliana* salt tolerance protein STO	Li et al., 2006
GmOSBP	May be involved in some physiological reactions for stress-response and cotyledon senescence in the soybean	Li et al., 2008
GmbZIP132	Induced by drought and high salt treatments, over expression in *Arabidopsis* could increase the salt tolerance during germination and the seedling stage	Liao et al., 2008
GmOLPa	The mature GmOLPa protein without the signal peptide has a calculated molecular mass of 21.5 kDa and a *pI* value of 4.4	Onishi et al., 2006

Aghaei et al. (2008) evaluated the changes in protein expression using the proteomic approach to assess the response of soybean to salt stress. Proteins from the hypocotyls and roots treated with 100 mmol L^{-1} NaCl were extracted and separated by using two-dimensional polyacrylamide gel electrophoresis. Some 321 protein spots were detected. In response to salt stress, seven proteins were reproducibly found to be up- or down-regulated by 2-fold to 7-fold as follows: late embryogenesis-abundant protein, β-conglycinin, elicitor peptide three precursor and basic/helix-loop-helix protein were up-regulated whereas protease inhibitor, lectin, and stem 31-kDa glycoprotein precursor were down-regulated. These results indicate that salinity can change the expression level of some special proteins in the hypocotyl and root of the soybean. This in turn may play a role in the soybean adaptation to saline conditions.

7. PERSPECTIVES OF BREEDING SOYBEAN FOR SALT TOLERANCE

7.1 Fine mapping of salt tolerance genes

To date limited progress in the generation of salt-tolerant soybean genotypes has been made through conventional breeding programs. In China, a few varieties such as Tiefeng 8 and

Zhonghuang 13 have been released. By using the salt-tolerant germplasm Wenfeng 7 as a parent, Shao developed a salt-tolerant cultivar Zhonghuang 10 that has been planted on 26,666.7 ha of saline lands. The evaluation and utilization of salt-tolerant soybean germplasm was awarded the Beijing Sci-Tech Progress Award (II) in 2002. Three populations of salt-tolerant soybean cultivars, which included the salt-tolerant Wenfeng 7 and sensitive cultivars were used to map the salt tolerance genes. A co-dominant RAPD marker was identified via analysis of the segregation of the F_2 plants (Guo et al., 2000). The marker, which was closely linked to the salt-tolerant/sensitive alleles is now being utilized to select salt-tolerant, high-yielding soybean lines. The identification and utilization of this marker obtained the Chinese patent in 2004. Zhang et al. (2005) evaluated Chinese soybean germplasm by using a SSR marker linked to a salt tolerance QTL (Lee et al., 2004). The 212 bp at SSR locus Satt339 in the salt-tolerant germplasm accounted for 74%, which indicated the potential usefulness of this marker in salt tolerance soybean breeding. Guo et al. (2000) reported the conversion of the RAPD marker into SCAR marker. These authors have recently developed a co-dominant SCAR marker. The identity of the genotype and phenotype was 85.1% when the marker was used to analyze 55 salt-tolerant and 90 salt-sensitive soybean germplasm (unpublished data). These results demonstrate that these markers will be useful in the identification of salt-tolerant soybean germplasm and marker-assisted selection of salt tolerance in soybean breeding.

However, the progress mentioned above has resulted from the limited understanding of the genetic and physiological mechanisms of salt tolerance. More focused efforts are needed to screen germplasm, determine new genes, develop useful molecular markers and gene-tagging methodologies, and to combine genes for higher levels of salt tolerance in soybean. Despite the efforts made on the molecular and physiological aspects of drought and salt tolerance, further studies are needed to better understand the mode of action for ion exclusion under salt stress. This will enhance the capability to develop salt-tolerant soybean cultivars. A significant number of QTLs/genes have been identified for different traits in soybean but only a few of them are associated with salt tolerance. More studies are needed to elucidate and determine novel genes and their mode of action for high salt tolerance. Although there are numerous cultivated and wild accessions in the soybean germplasm collections of the world, limited germplasm with salt tolerance has been further identified. Combining genes from both wild and cultivated species shows promise to obtain genotypes with enhanced levels of salt tolerance (Luo et al., 2005a). The mapping for new QTLs/genes and the determination of the gene-induced action under salt stress will provide the key resources to improve tolerance to drought and salt stresses in soybean.

7.2 Pyramiding genes by marker-assisted selection (MAS) in soybean breeding

Marker-assisted selection (MAS) will be important for pyramiding genes at two or more loci to enhance salt tolerance in soybean. The limitations of molecular markers have been overcome with the discovery and use of gene-based abundant SNP markers. SNP and other molecular markers have helped to develop a high-density soybean genetic map for the identification and characterization of QTLs/genes conditioning salt tolerance-related traits, thus facilitating MAS programs. The United States Department of Energy and the Joint Genome Institute (DOE-JGI) has released a preliminary whole soybean genome shotgun at the beginning of 2008 (http://www.phytozome.net/soybean.php). The availability of the soybean genome sequence information, the integrated soybean genetic and physical map and traits-specific SNP markers will make a significant contribution to the molecular breeding for abiotic stresses. The genetic engineering technology is an attractive approach to improving soybean for drought and salt stress tolerance. Some promising genes have been identified and are being used successfully to develop genetically engineered drought and salt-tolerant rice and canola. The introduction of drought and salt tolerance genes from soybean and other crops into elite soybeans that have high yield, enhanced resistance to pathogens and improved seed quality is highly desirable. Molecular techniques like QTL mapping, gene cloning, gene transformation and DNA microarray and gene expression analysis related to specific QTL regions are rapidly progressing. They will play a vital role in the development of stress-tolerant soybeans.

Effective use of the available genetic resources, a better understanding of the salt tolerance mechanisms, the construction of a fine map of the genome, the development of MAS techniques and the use of well-designed breeding strategies will contribute to advancing the development of soybean varieties with significantly greater salt tolerance.

REFERENCES

Abel, G. H., & MacKenzie, A. J. (1964). Salt tolerance of soybean varieties (*Glycine max* L. Merrill) during germination and later growth. *Crop Sci*, *4*, 157-161.

Abel, G. H. (1969). Inheritance of the capacity for chloride inclusion and chloride exclusion by soybeans. *Crop Sci*, *9*, 697-698.

Aghaei, K., et al. (2008). Proteome analysis of soybean hypocotyl and root under salt stress. *Amino Acids*, *36*(1). 91-98.

An, P., et al. (2002). Salt tolerance in two soybean cultivars. *J. Plant Nutr*, *25*(3), 407-423.

Aoki, A., et al. (2005). Molecular cloning and characterization of a novel soybean gene encoding a leucine-zipper-like protein induced to salt stress. *Gene*, *356*, 135-145.

Bernstein, L., & Ogata, G. (1966). Effects of salinity on nodulation, nitrogen fixation and growth of soybeans and alfalfa. *Agron. J*, *58*, 201-203.

Blumwald E., & Grover, A. (2006). Salt tolerance, (Chapter 11)., In Halford, N.G. (Eds.), *Plant Biotechnology: Current and future uses of genetically modified crops*, John Wiley and Sons Ltd, UK, 206-224.

Chang, R., et al. (1994). Effect of salt stress on agronomic characters and chemical quality of seeds in soybean. *Soybean Sci*, *13*(2), 101-105.

Chen, M., et al. (2007). GmDREB2, a soybean DRE-binding transcription factor, conferred drought and high-salt tolerance in transgenic plants. *Biochem. Biophys. Res. Commun*, *353*(2), 299-305.

Chinnusamy, V., et al. (2005). Understanding and improving salt tolerance in plants, *Crop Sci*, *45*, 437-448.

El-Samad, H. M., & Shaddad, M. A. K. (1997). Salt tolerance of soybean cultivars. *Biol. Plant.* *39*(2), 263-269.

Epstein, E., et al. (1990). Saline culture of crops: A general approach. *Science*, *210*, 399-404.

Essa, T. A. (2002). Effect of salinity stress on growth and nutrient composition of three soybean (*Glycine max* L. Merrill) cultivars, *J. Agron. Crop Sci*, *188*(2), 86-93.

Gao, J., & Lin, H. (2005). Review of salt-tolerant mechanism on rice-a research on salt-tolerant quantitative trait locus (QTL) *SKC1*. *Chin. Sci. Bull*, *17*, 563-565.

Guo, B., et al. (2000). Tagging a salt-tolerant gene using PCR markers in soybean. *Sci. Agric. Sin*, *33*(1), 10-16.

Greenway, H., & Munns, R. (1980). Mechanisms of salt tolerance in non-halophytes. *Ann. Rev. Plant Physiol.*, *31*, 149-190.

Hoagland, D. R., & Arnon, D. I. (1938). The water-culture method for growing plants without soil. *Calif. Agric. Expt. Circ*, *347*, 1-39.

Johnson, C. M., et al. (1957). Comparative chlorine requirements of different plant species. *Plant Soil*, *8*, 337-353.

Joly, R. J. (1989). Effect of sodium chloride on the hydraulic conductivity of soybean root systems. Plant Physiol, *91*(4), 1262-1265.

Kao, W. Y., et al. (2006). Response of three *Glycine* species to salt stress. *Environ. Expt. Bot*, *56*(1), 120-125.

Katerji, N., et al. (2003). Salinity effect on crop development and yield, analysis of salt tolerance according to several classification methods. *Agric. Water Management*, *62*(1), 37-66.

Lauchli, A. (1984). Salt exclusion: an adaptation of legumes for crops and pastures under saline conditions, In Staples, R. C., & Toeniessen, G. H. (Eds.), *Salinity tolerance in plants:*

Strategies for crop improvement, Wiley, New York, 171-187.

Lauchli, A., & Wieneke, J. (1979). Studies on growth and distribution of Na^+, K^+, and Cl^- in soybean varieties differing in salt tolerance. *Zeitschrift Fur Pflanzenernahrung und Bodenkunde*, *142*, 3-13.

Lee, G., et al. (2004). A major QTL conditioning salt tolerance in S-100 soybean and descendent cultivars. *Theor. Appl. Genet*, *109*(8), 1610-1619.

Li, D., et al. (2008). Molecular characterization of a novel salt-inducible gene for an *OSBP* (oxysterol-binding protein)-homologue from soybean. *Gene*, *407*(1-2), 12-20.

Li, F., et al. (2006). Cloning and characterization of a salt-tolerance related gene from *Glycine max*, *Mol. Plant Breed*, *4*(4), 464-468.

Li, X. P., et al. (2005). Soybean DRE-binding transcription factors that are responsive to abiotic stresses. *Theor Appl. Genet*, *110*(8), 1355-1362.

Liao, H., et al. (2003). *GmPAP3*, a novel purple acid phosphatase-like gene in soybean induced by NaCl stress but not phosphorus deficiency. *Gene*, *318*, 103-111.

Liao, Y., et al. (2008). Role of soybean *GmbZIP132* under abscisic acid and salt stresses. *J. Integr. Plant Biol*, *50*(2), 221-230.

Liu, T., et al. (2000). Salinity induced nuclear and DNA degradation in meristematic cells of soybean (*Glycine max* L.) roots. *Plant Growth Regul*, *30*(1), 49-54.

Luo, G., et al. (2005b). A putative plasma membrane cation/proton antiporter from soybean confers salt tolerance in *Arabidopsis*. *Plant Mol. Biol*, *59*(5), 809-820.

Luo, Q., et al. (2005a). Differential sensitivity to chloride and sodium ions in seedlings of *Glycine max* and *G. soja* under NaCl stress. *J. Plant Physiol*, *162*(9), 1003-1012.

Ma, S., & Wang W. (1994). Research of salt tolerance of soybean germplasm. *J. Jilin Agric. Sci*, (4), 69-71.

Onishi M., et al. (2006). Molecular cloning and characterization of a novel salt-inducible gene encoding an acidic isoform of PR-5 protein in soybean [*Glycine max* (L.) Merr.]. *Plant Physiol. Biochem*, *44*(10), 574-580.

Pan, R., et al. (2007). Identification of salt-tolerant soybean at sea shore in Hebei province, *Anhui Agri. Sci. Bull*, *13*(7), 158-159.

Pantalone, V. R., et al. (1997). Chloride tolerance in soybean and perennial *Glycine* accessions. *Euphytica*, *97*(2), 235-239.

Parker, M. B., et al. (1986). *Sensitivity of soybean cultivars to soil chloride*, Research Bulletins 347, The Georgia Agricultural Experiment Stations, University of Georgia, USA, 1-14.

Serraj, R., et al. (1994). Salt stress induces a decrease in the oxygen uptake of soybean nodules and in their permeability to oxygen diffusion. *Physiol. Plant*, *91*(2), 161-168.

Shao, G. H., et al. (1986). Preliminary studies on the evaluation of salt tolerance in soybean varieties. *Sci. Agric. Sin*, *6*, 30-35.

Shao, G. H. (1988). Relationship between distribution of salt-tolerant soybean varieties and saline soil. *Crops*, *2*, 26.

Shao, G. H., et al. (1994). Study on inheritance of salt tolerance in soybean. *Acta Agron. Sin*, *20*(6), 721-726.

Shao, G. H., et al. (1995). Screening for salt tolerance to soybean cultivars of the United States. *Soybean Genet. Newsl*, *22*, 32-42.

Stoddard, F. L., et al. (2006). Screening techniques and sources of resistance to abiotic stresses in cool-season food legumes. *Euphytica*, *147*(1-2), 167-186.

Sun, Y., et al. (2006). Studies on the over expression of the soybean *GmNHX1* in *Lotus corniculatus*: The reduced Na^+ level is the basis of the increased salt tolerance. *Chin. Sci. Bull*, *51*(11), 1306-1315.

Umezawa, T., et al. (2000). Enhancement of salt tolerance in soybean with NaCl pre-treatment. *Physiol. Plant*, *110* (11), 59-63.

Umezawa, T., et al. (2002). Discrimination of genes expressed in response to the ionic or osmotic effect of salt stress in soybean with cDNA-AFLP. *Plant Cell Environ*, *25*(12), 1617-1625.

Yang, J., & Blanchar, R. W. (1993). Differentiating chloride susceptibility in soybean cultivars. *Agron. J*, *85*, 880-885.

Zhang, Q., et al. (1999). *RAPD markers associated with salt tolerance in wild soybean populations*, Soybean Genetics Newsl., http://www.soygenetics.org/articles/sgn1999-010.html.

Zhang, H., et al. (2005). Genetic diversity analysis and marker assisted identification of salt-tolerant soybean by using SSR marker. *J. Plant Genet. Resour*, *6*(3), 251-255.

Zhong, M., et al. (1997). Search for molecular markers of salt tolerance of soybean by DNA amplification fingerprinting. *Soybean Genetics Newsl*, *24*, 81-82.

CHAPTER 8.

CONCLUSIONS AND RECOMMENDATIONS

Luxiang Liu*, Ahsan Irshad and Jiayu Gu

National Key Facility for Crop Gene Resources and Genetic Improvement, Institute of Crop Sciences, Chinese Academy of Agricultural Sciences, Beijing 100081, China

*Correspondence: luxiang@263.net.cn

More than 6% of the world's total land area is affected by saline stress, and a high proportion of cultivated land is increasingly becoming salinated due to clearcutting, excessive irrigation, and factors related to climate change such as salt intrusion into coastal zones as a result of sea-level rise and more frequent storms (Smajgle et al., 2015; Acosta-Motos et al., 2017). Excessive levels of salt in soil can negatively affect plant growth and productivity, substantially reducing food production for major crops such as rice and wheat (Acosta-Motos et al., 2017). The cereals which are major agricultural products mainly produced from glycophytes are highly sensitive to salt stress. Despite considerable progress in our understanding of how plants respond to salt stress, the breeding of salt-tolerant crops and cultivars is slow, with only limited progress with rice, soybean and wheat. This chapter covers research progress related to biotechnological approaches, the molecular processes involved in salt-stress responses in cereals, and how these processes are regulated. We will highlight evolutionary traits that plants have evolved to defend against salt accumulation in plant tissues and how to reduce the impact of salt accumulation on growth and productivity.

1. MOLECULAR AND BIOTECHNOLOGICAL APPROACHES

1.1 RNA interference (RNAi) technologies for the production of salt-tolerant plants

The biological process in which small RNA (sRNA) molecules stop the process of translation of targeted mRNA and alter the expression of the gene is known as RNAi. In plants, it is an important mechanism which is taking part in the post-transcriptional regulation of gene expression (Kumar et al., 2017). A kind of complex non-coding RNA such as microRNAs (miRNAs) and small interfering RNAs in plants has been evolved to silence the target gene without affecting the other genes. It is a powerful approach for silencing the target genes or changing the expression of gene for the improvement of cereal crops. With the help of genomic tools such as degradome analysis and deep sequencing, it has been identified that an increasing number of miRNAs are mainly involved in the growth and developmental processes. They have also facilitated the identification and analysis of stress responses in plants. Transgenic plants displayed a germination rate of 73%–78%, with 72%–77% of seedlings turning green at levels of 150 mmol L^{-1} NaCl. In contrast, wild-type plants displayed only a germination rate of 63% and a greening rate of 57.6% at the same NaCl level (Li et al., 2015). The inhibition of chlorophyll biosynthesis occurs due to salt stress. Pan et al. (2016) determined that *miR172a* plays an enhanced role in salinity tolerance and long-distance stress signaling in soybeans. The investigation was done by checking the growth of *miR172a* hairy root lines and *miR172a* deficient lines which were under the stress of 200 mmol L^{-1} NaCl of salt stress.

1.2 QTLomics of salt tolerance

The identification and analysis of QTLs are now used for marker development and crop breeding (Cobb et al., 2019). Thanks to extensive physiological and molecular studies assessing tolerance to various abiotic factors (ionic and/or osmotic stresses) and an information gateway on stress-specific and shared stress adaptation mechanisms, it is now possible to characterize genes associated with abiotic stress tolerance at the molecular level, especially those related to salt tolerance. The successful identification of QTLs in cereal crops such as rice, wheat and maize also helps to improve the crop against salt stress. These significant findings of QTLs, especially *Saltol* QTL, are mainly associated with Na^+/K^+ homeostasis under salt stress (Thomson et al. 2010).

1.3 Exploring halobiomes as gene pools for the production of salt-tolerant plants

A group of halophilic and/or halotolerant plants, bacteria and fungi which have the ability to withstand the hypersaline environment are known as halobiomes. These are potential reservoirs of the genes which are associated with salt tolerance and help to improve salt-tolerant crops. Salt tolerance in halophytes is due to structural adaptations which include leaf shedding and salt secreting assemblies (salt glands or salt hairs) (Shabala et al., 2014). Halophytes achieve their tolerance by cellular mechanisms with reduction of sodium influx or elimination of Na^+ with the help of plasma membrane anti-porters (Flowers and Colmer, 2015). The mechanism of salt tolerance expression is higher in halophytes as compared to glycophytes.

The quantitative differences are more important as compared to qualitative differences in the expression of different genes and their active proteins which are involved in salt tolerance for survival of the halophytes (Himabindu et al., 2016). Different studies have evaluated that genes such as *SOS, NHX*, and *HKT* provided better tolerance in halophytes than those isolated from glycophytes (Volkov, 2015). So, it is important to select the effective promoters and regulatory sequences for the achievement of a proper level of transgene expression. It has been demonstrated in recent studies that different cis-regulatory and promoter regions of different stress-reduced genes have been found in the halophytes which show high expression under salinity conditions. In some plants, genes showed the multi-fold up-regulation of their associated genes in salt stress such as in *Suaeda liaotungensis*, the promoter region of *SIBADH* and *SIPEAMT* genes; in *T. halophila*, the *TsVPI* gene and the *SbGSTU* gene of *Salicornia brachiate* (Li et al., 2016; Tiwari et al., 2016). Similarly, salt tolerance was enhanced in glycophytes by introducing the genes from halophytes including transcription factors (*MYB, NAC, BRED*) which were associated with enzymatic or non-enzymatic antioxidants and antiporter genes (*NHX, HKT, SOS*) (Geilfus et al., 2017).

1.4 Plant genome editing for producing salt-tolerant plants

The breeding of various crops has been revolutionized by specific and predictable modifications of the genomes through genome editing (GE) methods (van der Wiel et al., 2017). Customized nucleases are used for the introduction of mutation such as insertion or deletion or changing in the gene sequences for GE methods (Jain, 2015). Different methods of GE are now readily available and used in different crops such as *ZFNs* (zinc-finger nucleases), CRISPR-Cas9 (cluster regularly interspaced short palindromic repeat associated nuclease 9) and *TALENs* (transcriptional activator-like effector nucleases) (Kumar and Jain, 2015). The double-strand breaks (DSBs) induced by these sequence-specific nucleases help to break the DNA at specific points in the genome. This DNA is subsequently followed by DNA repair process at the same time by resulting genomic modifications such as insertion or gene replacement (Khan et al., 2017)

In several important crops, GE technology has been successfully used but little work has been done to develop salt-tolerant plants through this technology (Puchta, 2017). The main reason is that there is a need to manipulate multiple genes and due to complexity of the salt-tolerant traits and also their multiple pathways, there is a need to introduce site-directed changes with no effect on other yield related traits (Shi et al., 2017). The genomic sequence of maize was edited in gene

ARGOS8 by using CRIPS-Cas method and these mutant lines showed higher expression in most of the tissues and growth phases (Zörb et al., 2015). So it has been suggested that this method can be used in different crops by editing specific genes to develop salt-tolerant plants with high grain yield (Shi et al., 2017). Similarly *SAPK2* gene was mutagenized by GE which plays a role in ABA signaling osmotic stress and the mutant line had ABA-insensitive phenotype (Lou et al., 2017). On the basis of these results, it can be said that this gene can be used to improve salt and osmotic stress tolerance. GE can also be used to study the modification of genes on functionality of plants and help in the formation of specific functional plants.

2. PROGRESS IN BREEDING SALT-TOLERANT CEREAL CROPS

Different plant responses under salt stress conditions which affect plant growth, yield and quality of the crop dramatically slow the breeding program for salt-tolerant cultivars by using the standard method which mostly depends on the phenotypic selection. But there is still have sufficient genetic diversity to develop the tolerant varieties by identifying the functional alleles from the major genes which showed tolerance for major traits and combine these alleles in high-yielding backgrounds (Zörb et al., 2019). There is observed intrinsic diversity in different germplasm of crops which has been exploited through breeding to develop salt-tolerant varieties, but there is still not much progress has been done to develop commercial salt-tolerant varieties in rice and wheat (Shahbaz and Ashraf, 2013). Similarly, there is a need to understand the physiology and genetics of other crops for making salt-tolerant varieties such as soybean, maize and improvement of crop production (Guan et al., 2014). This study provides examples from three important cereal crops (rice, wheat and barley), whose salt stress responses are relatively well understood and in which progress has been made in breeding salt-tolerant cultivars (particularly in rice and wheat).

2.1 Progress in developing salt-tolerant rice

Rice is the most important staple crop in the world which have ability to grow in a wide range of ecological environments. Rice can be grown in dry uplands area and also in wet lowlands. Similarly, it can be cultivated in flooded soils and wide geographic ranges such as from 50°N in northern China to 35°N in South Australia (Li et al., 2017). Due to this rice has significant genetic diversity, it can grow in extreme geographic and ecological conditions and it is also suitable for genetic studies for improvement in the adverse condition of weather and soil conditions. Different studies in rice helped to identify QTLs which were associated with salt tolerance and had major effects on crops (Thomson et al., 2012; Tiwari et al., 2016). The major *Saltol*

QTL was identified on chromosome 1 in the Pokkali landrace which was involved in Na^+/K^+ ratios homeostasis under salt stress (Negra et al., 2011). Similarly, *SKC1* QTL was identified that it was derived from Nona Bokra landrace and was associated with shoot K^+ concentration. Later, it was identified as *OsHKT1;5* (Kaleem et al., 2018). These QTLs were recently fine mapped and markers were developed and transferred into different varieties through marker-assisted breeding (Linh et al., 2012).

2.2 Progress in developing salt-tolerant wheat

Wheat is more salt-sensitive as compared to barley, but it shows more resistance as compared to other cereal crops such as maize, rice and sorghum. Different studies have been reported related to genetic variations in the salt tolerance of wheat but the breeding of the salt tolerance cultivars is less and due to this considerable loss in the production of wheat was inflicted by salt stress (Ismail et al., 2017). Due to conventional breeding in India, Pakistan and China, a little progress has been done in the development of new varieties, such as the development of S-24 variety released in Pakistan which has considerable amount of high yield and good quality (Shahbaz and Ashraf, 2013).

Using the functional alleles of *Nax1*, *Nax2* and *Kna1* in the high-yielding cultivars will help to improve the salt tolerance in bread and durum wheat. Similarly, these QTLs can also help to improve the other cereal crops' salt-resistance by using them through transformation or GE with knowing that it will not affect the quality and yield of the crops. In the future, improvement of these crops related to salt tolerance will be a challenge due to massive use of land which is progressively deteriorating, use of new land for crop production, and increasement in the demand of food due to the population in the world (Tiwari et al., 2016). So, there is a need to conduct more studies on these crops especially in bread and durum wheat to find out more loci and genes associated with salt tolerance and other important traits thus it can be combined in new genotypes by new molecular breeding tools.

2.3 Progress in developing salt-tolerant barley

Barley is the most salt-tolerant crop among glycophytes and it can be used as a model crop for improving the salt tolerance of cereal crops. There are considerable genetic variabilities for salt tolerance in cultivated and wild relatives of barley (Zhu et al., 2015). These genetic variabilities provide the opportunity to develop new salt-tolerant varieties. But still little progress has been done for variety development in barley. This is likely due to barley's inherent salt stress tolerance compared to other cereals. Many traits in barley are associated with salt tolerance such as salt exclusion, high K^+ concentration in leaf, accumulation of inorganic ions in the shoot due to osmotic adjustment and higher stomatal density (Zhu et al., 2017). The over-expression of *HvHKT2;1*, which is the transporter permeable to both K^+ and Na^+, helps to enhance the Na^+ uptake with translocation to the shoot under salt stress and also increased salt tolerance in barley (Mian et al., 2011). Similarly, in recent studies, it has been reported that high K^+/Na^+ ratio levels and low concentration of Na^+ are associated with salt tolerance under mild stress in barley (Zhu et al., 2017). These different results give the suggestion that these responses are likely genotypic specific and for conclusive results, there is a need to study a larger set of diverse genotypes. Several salt-tolerant QTLs have been identified in barley and these QTLs further can be used to develop effective breeding approaches by using molecular markers (Sbei et al., 2014).

3. FUTURE PROSPECTS RELATED TO EMERGING TECHNOLOGIES

In cereal crops, little effort has been done for the development of salt-tolerant new cultivars and less work have been done related to understanding biological mechanism under salt stress in plants (Gilliham et al., 2017). Similarly, QTLs and GWAS studies that hoped to discover new loci relating to salinity tolerance for subsequent use in varietal development were not realized because most QTL studies did not pursue gene discovery and were not used for breeding, except for the *Nax1* and *Nax2* genes. It is predicted that recent advances in genomics and advancement in breeding will play crucial roles in basic and applied research to find out salinity tolerance in cereal crops. Recently new breeding technologies have been outlined by scientists which help to exploit the cereal genetic resources and these can be applied to improve salinity tolerance in cereals (Rasheed et al., 2018). It is important to get the genome sequence of bread wheat and also its relatives such as *A. taushii* for genetic analysis (Avni et al., 2017). By improving the genome, it will help to add values by improving the continuity and help to anchor the chromosomes as well as help to assist in breeding by integrating the genome sequence with linkage maps. Due to this, we can identify the favorable recombination in the genetic stock which was difficult before (Rasheed et al., 2018). With the help of whole genome sequencing, new genes have been identified in wild soybean (*GmCHX1*) and barley (Qi et al., 2014). These results indicated that whole genome resequencing is effective to identify new salt-tolerant genes in cereal crops. Whole genome sequencing is still an expensive technology and there is a need to use an available sequence of cultivated plants by using new and improved genotyping platforms for gene discovery. These platform of genomic sequence could be useful and also cost-effective with making it possible to apply for genomic selection and breeding strategies thus can improve the quantitative and complex traits with a high degree of accuracy (Hickey et al., 2017). A large population of genotypes (training population) can be accurately phenotyped and genotyped during genomic selection. The phenotype traits of population are regressed against markers and a prediction equation is generated for the traits from these markers. Selection candidates, which are not part of the training population, are genotyped and selected based on their predicted trait values. Crossing can be done between the selected candidates and the progeny from this cross can immediately be genotyped to facilitate the selection between them. Due to genomic selection, high intensity selection is enabled (due to the large population can be grown and genotyping cost is lower than when phenotype required) and cycle time is rapid (because selected candidates do not need to be phenotyped) (Mackay et al., 2015). It all depends upon the availability of high-throughput platform for salinity tolerance which totally reflects the performance of field. But still there is no any example for rice and wheat for genomic prediction studies. Therefore, there is a need to conduct these types of experiments soon because genomic selection can improve prediction accuracy at low costs. After the successful genomic selection in livestock breeding by getting high genetic gain and digitalization of germplasm with help of high-density finger printing, it can be said that genomic selection not only helps to identify loci of salt tolerance in GWAS but at the same time also identified salt-tolerant germplasm with desired genomic estimated breeding values under the condition of salt stress. Genomic estimated breeding values (GEBVs) are more accurate than breeding value estimates based on phenotype and pedigree alone (Hickey et al., 2017). The germplasm resources which are lying in the gene bank could be genotyped to narrow salt-tolerant genetic resource with the absence of field trials after optimizing the GEBVs for salt tolerance. In this way genomic selection could significantly increase the intensity and selection accuracy and also help to reduce the generation intervals. For attaining the rate of genetic gain, these attempts are very important in consideration. Speed breeding is a new optimized approach in cereal crops, which significantly decreases the in generation turnover with an increment in the rate of gene deployment (Watson et al., 2018). Speed breeding, international collaborations, effectively mining genetic resources in global gene banks and genomic selection (SpeedGS) will all accelerate the rate of genetic gain when improving locally-adapted salt-tolerant varieties of cereal crops in different regions (irrigated and rainfed).

4. CONCLUSION

A complex abiotic factor that mainly affects irrigated and non-irrigated land is known as salinity. In last few decades a considerable research and breeding efforts have been done in wheat, rice and maize to improve salinity tolerance. These often involved measuring and identifying single traits and underlying genes as surrogates that equate to salinity tolerance under field conditions. Despite extensive QTL studies assessing salt tolerance in cereals, identification of candidates and functional genes has progressed slowly and our knowledge is limited to a few functional genes. Improving the technology of genetics and breeding will help to fast the process of research on salt tolerance and innovation of new high-yield salt-tolerant varieties available to the farmers. It is important to identify high tolerant variants such as highly tolerant landraces and to identify salt-tolerant genes with yield advantages in saline conditions. So, there is a need to do breeding/selection for increasing the salt-tolerant derivatives, just like breeding/selection used for drought tolerance where different generations show tolerant in drought conditions. There is a need to use genetic resources carefully because salt tolerance is a complex and polygenic trait. Careful attention should be needed during screening in field trial in saline condition with checking the plant response and focusing on the yield performance. It's not a feasible approach that just focusing on salt tolerance in cereal crops during breeding. Breeders should develop high-yielding varieties under stress and non-stress conditions in the future. There is a need to use rapid generation breeding techniques such as double haploid which can manipulate the chromosome recombination and pairing and also improve the gene editing techniques.

There is a need to design and evaluate carefully the transcription factors or other factors associated with salt stress such that they do not describe and target less relevant factors associated with salt stress response. It is also necessary to know the salt tolerance process that occurs inside the plants where salt concentrations are increasing with passage of time. The tolerance of transcription factor and genes affected by salt shocks depend on variation of salt level in the field. So, on the basis of these results breeding of salt-tolerant crops can be done. The breeding on the basis of the seedling stage results for salt tolerance may be less relevant according to the agriculture point of view. Therefore, it is important to do field trials with specific control to evaluate the yield and also improve crop productivity (Roy et al., 2014).

Different mechanisms have been developed by land plants for salt tolerance. For example, different plants have different gene expression patterns, different transporters and also ion flux regulation mechanisms. In cereal crops, there is still limited information about ions transporters which are involved in the uptake, sequestration, exclusion and salt transportation. Additionally, there are several uncertainties, nomenclatural problems, and contradictions based on hastily-made generalized statements, leading to several unanswered questions about salt tolerance. The rules of salt tolerance mechanisms are few and salt tolerance mechanisms vary from species to species, population to population, cultivar to cultivar and between different organ and tissues (Nieves-Cordones et al., 2016). Different types of plant experiments in field and control conditions help to understand the impact of salt stress and how plants respond to salt. The same ion transporter can affect the shoot Na^+ accumulation when modified (down- or up-regulated) in different cell types (e.g. root epidermal cells vs. xylem cells), meaning the expression of ion transporter/transport components could be regulated at the cell-specific level. Little differences in protein/gene or transcription factor structure have significant effects on plant's ability to control ion fluxes which help to tolerate salt stress. It is suggested that there is a need for further functional analysis and characterization of various ions transporters in different organisms. Similarly, more focus is needed on tissue and cell expression patterns and also on the characterization of the halophytes which are the potential source of tolerant gene or transcription factors and can help to improve tolerance in glycophytes (Mishra and Tanna, 2017). Over-expression of Salsola soda *NHX* gene is an example of alfalfa plants in which plants can grow in 400 mmol L^{-1} NaCl over fifty days due to over-expression (Li et al., 2011). Improved ion sequestration in the vacuole under salt stress decreases the solute potential of the cells and alleviates the osmotic component of salt stress by facilitating water uptake. Engineering ion transport is a component which helps to improve salt tolerance. Improving salt tolerance is a demanding and complex task in the research and needs more focus especially when the main purpose is to develop a tolerant variety with high-yielding for commercial use. This is a significant challenge even when using the most recent genetic engineering tools or using transcription factors since alterations in ion homeostasis typically generate yield burden. Given all of this, it should be no surprise that genetically modified crops with increased salt tolerance have yet to be commercialized.

REFERENCES

Avni, R., et al. (2017). Wild emmer genome architecture and diversity elucidate wheat evolution and domestication. *Science*, *357*, 93-97.

Cobb, J. N., et al. (2019). Back to the future: revisiting MAS as a tool for modern plant breeding. *Theor. Appl. Genet*, *132*, 647.

Acosta-Motos, J. R., et al. (2017). Plant responses to salt stress: adaptive mechanisms. *Agronomy*, *7*, 18.

Flowers, T. J., & Colmer, T.D. (2015). Plant salt tolerance: adaptations in halophytes. *Ann. Bot*, *115*, 327-331.

Geilfus, C. M., et al. (2017). Transient alkalinization of the leaf apoplast stiffens the cell wall during onset of chloride salinity in corn leaves. *J. Biol. Chem*, *292*, 18800-18813.

Gilliham, M., et al. (2017). Translating knowledge about abiotic stress tolerance to breeding programmes. *Plant J*, *90*, 898-917.

Guan, R., et al. (2014). Salinity tolerance in soybean is modulated by natural variation in *GmSALT3*. *Plant J*, *80*, 937-50.

Hickey, J. M., et al. (2017). Genomic prediction unifies animal and plant breeding programs to form platforms for biological discovery. *Nat. Genet*, *49*, 1297-1303.

Himabindu, Y., et al. (2016). Salt-tolerant genes from halophytes are potential key players of salt tolerance in glycophytes. *Environ. Exp. Bot*, *124*, 39-63.

Ismail, A. M., & Horie, T. (2017). Genomics, physiology, and molecular breeding approaches for improving salt tolerance. *Ann. Rev. Plant Biol*, *68*, 405-434.

Jain, M. (2015). Functional genomics of abiotic stress tolerance in plants: a CRISPR approach, Front. Plant *Sci*, *6*, 375.

Kaleem, F., et al. (2018). An overview of the genetics of plant response to salt stress: present status and the way forward. *Appl. Biochem. Biotechnol*, *186*, 306-334.

Khan, A., et al. (2017). Nitrogen fertility and abiotic stresses management in cotton crop: a review. *Environ. Sci. Pollut. Res*, *24*, 14551-14566.

Kumar, V., & Jain, M. (2015). The CRISPR-Cas system for plant genome editing: advances and opportunities. *J. Exp. Bot*, *66*, 47-57.

Kumar, V. (2017). Plant small RNAs: the essential epigenetic regulators of gene expression for salinity stress responses and tolerance. *Plant Cell Rep*, *37*, 61-75.

Li, Q.L., et al. (2016). Molecular cloning of phosphoethanolamine *N*-methyltransferase, (PEAMT). gene and its promoter from the halophyte *Suaeda liaotungensis* and their response to salt stress. *Acta. Physiol. Plant*, *38*, 39.

Li, W., et al. (2015). Overexpression of soybean miR172c confers tolerance to water deficit and salt stress but increases ABA sensitivity in transgenic *Arabidopsis thaliana*. *J. Exp. Bot*, *67*, 175-194.

Li B., et al. (2017). Chloride on the move. *Trends Plant Sci*, *22*, 236-248.

Linh, L .H., et al. (2012). Molecular breeding to improve salt tolerance of rice (*Oryza sativa* L.) in the Red River Delta of Vietnam. *Int. J. Plant Genom*, 949038.

Lou, D., et al. (2017). *OsSAPK2* confers abscisic acid sensitivity and tolerance to drought stress in rice. *Front. Plant Sci*, *8*, 993.

Mackay, I., et al. (2015). GplusE: beyond genomic selection. *Food Energy Secur*, *4*, 25-35.

Mian, A., et al. (2011). Over-expression of an Na^+- and K^+-permeable *HKT* transporter in barley improves salt tolerance. *Plant J*, *68*, 468-79.

Mishra, A., & Tanna, B. (2017). Halophytes: potential resources for salt stress tolerance genes and promoters. *Front. Plant Sci*, *8*, 829.

Munns, R. (2006). Approaches to increasing the salt tolerance of wheat and other cereals. *J Exp Bot*, *57*, 1025-1043.

Negrão, S., et al. (2011). Recent updates on salinity stress in rice: from physiological to molecular responses. *Crit. Rev. Plant Sci*, *30*, 329-77.

Nieves-Cordones, M., et al. (2016). Comparison between *Arabidopsis* and rice for main pathways of K^+ and Na^+ uptake by roots. *Front. Plant Sci*, *7*, 992.

Pan, J. W. (2016). Soybean *miR172a* improves salt tolerance and can function as a long-distance signal. *Mol. Plant*, *9*, 1337-1340.

Puchta, H. (2017). Applying CRISPR/Cas for genome engineering in plants: the best is yet to come. *Curr. Opin. Plant Biol*, *36*, 1-8

Qi, X., et al. (2014). Identification of a novel salt tolerance gene in wild soybean by whole-genome sequencing. *Nat. Commun*, *5*, 1-11.

Rasheed, A., et al. (2018). Wheat genetic resources in the post-genomics era: promise and challenges. *Ann. Bot*, *121*, 603-616.

Ren, Z. H., et al. (2005). A rice quantitative trait locus for salt tolerance encodes a sodium transporter. *Nat. Genet*, *37*, 1141-46.

Roy, S. J., et al. (2014). Salt resistant crop plants. *Curr. Opin. Biotechnol*, *26*, 115-124.

Sbei, H., et al. (2014). Detection of QTLs for salt tolerance in Asian barley (*Hordeum vulgare* L.). *Breed Sci*, *64*, 378-88.

Shabala, S., et al. (2014). Salt bladders: do they matter?. *Trends Plant Sci*, *19*, 687-691.

Shi, J., et al. (2017). ARGOS8 variants generated by CRISPR/Cas9 improve maize grain yield under field drought stress conditions. *Plant Biotechnol. J*, *15*, 207-216.

Smajgle, A., et al. (2015). 2015. Responding to rising sea levels in the Mekong Delta. *Nat. Clim. Change*, *5*, 167-74.

Thomson, M. J., et al. (2010). Characterizing the Saltol quantitative trait locus for salinity tolerance in rice, *Rice*, *3*, 148-160.

Thomson, M. J., et al. (2012). High-throughput single nucleotide polymorphism genotyping for breeding applications in rice using the Bead Xpress platform. *Mol. Breed*, *29*, 875-86.

Tiwari, S., et al. (2016). Mapping QTLs for salt tolerance in rice (*Oryza sativa* L.) by bulk segregant analysis of recombinant inbred lines using 50K SNP chip. *PLoS one*, *11*(4), e0153610.

Tiwari, V. (2016). Functional characterization of the tau class glutathione-S-transferases gene, (*SbGSTU*). promoter of *Salicornia brachiata* under salinity and osmotic stress. *PLoS ONE*, *11*, e0148494.

van de Wiel, C. C. M., et al. (2017). New traits in crops produced by genome editing techniques based on deletions. *Plant Biotechnol. Rep*, *11*, 1-8.

Volkov, V. (2015). Salinity tolerance in plants. Quantitative approach to ion transport starting from halophytes and stepping to genetic and protein engineering for manipulating ion fluxes.

Front. Plant Sci, *6*, 873.

Watson, A., et al. (2018). Speed breeding is a powerful tool to accelerate crop research and breeding. *Nat. Plants*, *4*, 23-29.

Zhu, M., et al. (2017). Physiological and molecular mechanisms mediating xylem Na^+ loading in barley in the context of salinity stress tolerance, *Plant Cell Environ*, *40*(7), 1009-1020.

Zhu, M., et al. (2015). Linking osmotic adjustment and stomatal characteristics with salinity stress tolerance in contrasting barley accessions. *Funct. Plant Biol*, *42*, 252-63.

Zörb, C., et al. (2019). Salinity and crop yield. *Plant Biol*, *21*, 31-38.

Zörb, C., et al. (2015). Salinity stiffens the epidermal cell walls of salt-stressed maize leaves: is the epidermis growth- restricting?. *PLoS one*, *10*(3), e0118406.

PROTOCOL I

IN VITRO MUTAGENESIS PROTOCOLS FOR BREEDING SALT-TOLERANT WHEAT

Baocun Zhao, Rongchao Ge, Yinzhu Shen and Zhanjing Huang

College of Life Sciences, Hebei Normal University, Shijiazhuang 050016, China

1. INTRODUCTION

An excess accumulation of salt in the soil can cause salinization in the rhizosphere of cultivated crops. Salinization has become a main land degradation problem restricting agricultural production worldwide. Normal growth of crops will usually be affected when the soil salt content reaches 0.2%–0.5%. A higher soil salt content may cause wilting and even death of crops, resulting in a severe reduction of crop yield (Dai et al., 2003; Vinocur and Altman, 2005). With the continuous decline of available cultivated land due to extensive land degradation, people are increasingly paying attention to the utilization of saline-alkali lands. Practical experience has demonstrated that saline-alkali land can be utilized by breeding salt-tolerant varieties based on the difference and heredity of salt tolerance of plants and making selections of new varieties with better adaptation to the saline-alkali soil. Thus, by breeding salt-tolerant varieties of crops we can make use of the saline-alkali land and increase crop yield on that land. This work is of great relevance to the sustainable development of agriculture.

As plant cells have totipotency, a whole plant can be regenerated from a single cell. In using plant tissue culture somaclonal variations may occur after calli are induced. These somaclonal variations can provide a large number of usable resources for germplasm enhancement of crops. Mutation breeding is a modern breeding technology that has been increasingly used in the last 50 years. Its rapid development has produced a great impact on *in vitro* selection of salt-tolerant mutants. The *in vitro* mutation techniques combine *in vitro* culture and mutagenesis techniques to induce and select tolerant mutants in the process of plant tissue and cell culture. Research has shown that the combined use of *in vitro* culture and mutation induction can significantly enrich the types of mutations, increase mutation frequency and the overall breeding efficiency (Li et al., 1992).

Wheat (*Triticum aestivum* L.) is a very important cereal food crop. However, it is a glycophyte and has poor tolerance to salt. Therefore, breeding salt-tolerant wheat is especially important for agricultural production. Shimada et al. (1969) first induced calli from young embryos of wheat and then successfully obtained regenerated plants. Thereafter there have been more and more studies on wheat tissue culture. Wheat anther culture was first reported a success in China in the early 1970s (Zhu et al., 1973; Ouyang et al., 1973). Further studies focused on the application of the wheat anther culture in wheat breeding and a batch of new wheat varieties has been developed (Hu et al., 1983; de Buyser et al., 1987) demonstrating the use of the somaclonal variation in wheat breeding and agricultural production.

The advantages of plant tissue culture in the selection of a salt-tolerant mutant were discussed for the first time in 1972 (Melchers, 1972). In 1974, a salt-tolerant cell line was obtained using this technique (Zenk, 1974). Thereafter, the *in vitro* mutation technique has been widely used and played a very important role in plant resistance breeding.

The objective of this protocol is to describe the experimental procedures of the *in vitro* mutation technique and to provide an assessment of its efficiency and application status in breeding salt-tolerant wheat.

2. EXPERIMENTAL PROCEDURES

The breeding process of salt-tolerant wheat by *in vitro* (tissue and cell culture) mutagenesis includes donor plant material selection/determination, callus and mutation induction, mutant selection and identification. The experiments described herein used both wheat mature embryos and anthers as explants to conduct *in vitro* culture under appropriate conditions. The media composition for tissue culture is given in Appendix I. Calli were used to induce mutation and select salt-tolerant mutants. Salt-tolerant wheat lines were obtained from both explants of mature embryos and anthers. In continuation the detailed experimental procedures are described.

2.1 Donor plants and explants

The wheat varieties that donated explants were carefully chosen from local commercial wheat varieties with good agronomic characters. The tissues and organs used for explants were mature embryos and anthers. The mature embryos were obtained from the seeds of the above mentioned wheat varieties while the anthers were from F_1 generation of different cross combinations among these wheat varieties.

2.2 Callus induction and salt-tolerant mutant selection in mature embryo culture

2.2.1 Establishment of mature embryo culture

Wheat seeds were first disinfected in 0.1% $HgCl_2$ for 10 min, rinsed 3 times in sterile water and then disinfected in 75% ethanol for 5 min, rinsed 3 times in sterile water again. Treated seeds were soaked in sterile water and placed in a temperature box at 25℃ overnight. The mature embryos with scutella were excised in a laminar airflow transfer hood. After being dried on the sterile filter paper, the mature embryos were placed on the surface of the induction media with the scutella facing upwards to initiate the culture. The Murashige-Skoog (MS) basal medium supplemented with different concentrations of mutagens and hormones was used for callus and mutation induction. The mutagens Pingyangmycin (PYM) and Daunomycin or Zhengdingmycin (ZDM) were added at 0.1 $\mu g\ mL^{-1}$, 0.5 $\mu g\ mL^{-1}$ and 1.0 $\mu g\ mL^{-1}$ to the induction media to determine the optimum concentration for mutation induction. When putting the mature embryos on the media, the scutella must face upwards so as to prevent the germination of the mature embryos (Figure P1.1 in Appendix Ⅱ).

2.2.2 Callus induction

The mature embryos were cultured in a room at the temperature of 25–28℃. A total of 10–12 h of daily light and 1,500 lux light irradiance were provided for callus induction. Under these conditions, the calli formed after 3–4 days of culture (Figure P1.2 in Appendix Ⅱ). The callus induction frequencies were investigated according to different treatments and were calculated as follows :

$$\text{Frequency of callus induction} = \frac{\text{No. of mature embryos that produced calli}}{\text{No. of mature embryos placed on induction medium}} \times 100 \quad (\text{P1.1})$$

2.2.3 Basic experiment for salt tolerance

Disinfected seeds were put on the medium MS supplemented with 0.3%, 0.5% and 1.0% NaCl, respectively. Seed germination on the different media was observed. The NaCl concentration at which the seed germination percentage was 50% was used for the selection of salt-tolerant calli.

2.2.4 Calli selection for salt tolerance

The calli were transferred to MS based subculture media supplemented with 0.5% and 1.2% NaCl, respectively at the same time for subculture and salt tolerance selection. The salt concentration was not raised gradually in salt tolerance selection, thus effectively avoiding a false positive result if the calli had gained salt tolerance because of physiological adaptation. After 40–45 days of subculture, the number of surviving calli was counted. All those calli that were able to keep growing or produce fresh calli even if they were browning were counted as live ones. The survival rates of calli were calculated as follows:

$$\text{Survival rate of calli} = \frac{\text{No. of live calli}}{\text{No. of calli transferred on subculture and selection medium}} \times 100 \qquad (\text{P1.2})$$

2.2.5 Differentiation of regenerated plantlets

The surviving calli were transferred to 1/2 MS based differentiation media supplemented with 0.5%, 0.8% and 1.2% of NaCl respectively to promote the selected salt-tolerant calli to differentiate and generate new plantlets (Figure P1.3 in Appendix Ⅱ).

2.2.6 Transplanting

The differentiation status was investigated after 25–30 days of culture. The salt-tolerant plantlets were transplanted to normal soil (salt content lower than 0.3%) (Figure P1.4 in Appendix Ⅱ). Variations in the agronomic characters such as plant height, spike length, seed setting rate and grain plumpness were observed. The variant plants were classified.

2.2.7 Identification of salt tolerance

The stable individual mutant plants or mutant lines obtained in the M_2 generation were identified for their salt tolerance. Salt tolerance identification included the following.

2.2.7.1 Evaluation trial in an artificial salt pond

The obtained stable lines, together with the control variety (original wheat variety), were sown, and grown until the seeds were ripe in an artificial salt pond (soil containing 0.45% salt concentration) (Figure P1.5 in Appendix Ⅱ). The mature plants were harvested and their agronomic traits were inspected separately. The salt tolerance of the new lines was evaluated based on plant characters.

The salt tolerance efficiency and salt tolerance index were analyzed, thus determining the salt tolerance grade of the plant. The salt tolerance efficiency and salt tolerance index were calculated as follows:

$$\text{Salt tolerance efficiency} = \frac{\text{No. of survival material in salt treatment}}{\text{No. of survival material in control treatment}} \quad \text{(P1.3)}$$

$$\text{Salt tolerance index} = \frac{\text{salt tolerant efficiency of the new bred variety}}{\text{salt tolerant efficiency of the control variety}} \quad \text{(P1.4)}$$

The corresponding relation between the grade of salt tolerance and salt tolerance index were: Grade 1 $\geqslant$ 1.3, Grade 2 $\geqslant$ 1.1, Grade 3 $\geqslant$ 0.8, Grade 4 $\geqslant$ 0.5, and Grade 5 $<$ 0.5.

To determine the salt tolerance index, 3 replications of each material and 50 plants of every replication were investigated.

2.2.7.2 Physiological, biochemical and molecular studies

The stable lines were used to further explore the molecular mechanisms of salt tolerance (Shen et al., 1993; Bi et al., 1999; Suo et al., 2001; Wang et al., 2001; Wang et al., 2004).

2.3 Callus induction and salt-tolerant mutant selection in anther culture

2.3.1 Spikes sampling

The developmental stage of the microspores in the F_1 generation of different cross combinations was examined. After staining and slice production, the pollen grains of the F_1 generation were observed under the microscope. Anthers with the pollen grains at the middle-uninuclear stage were ready for *in vitro* culture. When sampling in the field, plants with mid-uninucleate pollens were chosen, the young spikes together with the flag leaves were picked up and put in polyethylene bags to prevent moisture loss. Samples were then pre-treated in the refrigerator at 4℃ for 72 h in order to increase the callus induction frequency.

2.3.2 Disinfection of young spikes

First the flag leaf and the base of the young spike were cut off. Then the leaf sheath was wiped with 75% ethanol. In the laminar airflow transfer hood, young spikes were shelled and put in 0.1% $HgCl_2$ to be disinfected for 10 min, then rinsed 3 times in sterile water and dried on sterile filter paper. The anthers in the middle spikelets were taken off and cultured on the N6 medium. The number of anthers placed on the medium depends on the size of the container. Usually it is appropriate to spread the anthers covering the whole surface of the medium.

2.3.3 Anther culture conditions

The anthers were cultured in an incubator at 27℃ to induce calli. A 12/12 (light/dark) h photoperiod was provided. Weak scattered light at the intensity of 15 dμE m^{-2} s^{-1} was adequate for callus induction. Although dark culture is beneficial to callus formation, it is easier to produce regenerated albino plantlets later. After 30–35 days, the callus induction frequency was investigated.

$$\text{Frequency of callus induction} = \frac{\text{No. of anthers that produced calli}}{\text{No. of anthers placed on induction medium}} \times 100 \qquad (P1.5)$$

2.3.4 Mutation induction treatment

Under aseptic conditions, the obtained calli were cut into small pieces about 2 mm^3 each, immersed in 0.4% EMS mutagen solution for 2 h, rinsed 3 times with sterile water, and then transferred to the N6 subculture medium. After 40–45 days, the surviving calli were counted and the survival rate was calculated. The criterion for live calli and the calculation formula for the survival rate of calli were the same as mentioned above.

2.3.5 Selection for salt tolerance

After being cut into small pieces, the surviving calli were transferred to the N6 medium supplemented with 0.5% NaCl for subculture and salt tolerance selection. The number of surviving calli was counted after 37 days of subculture. The survival rate of calli was calculated.

2.3.6 Differentiation of regenerated plantlets

The selected calli were transferred to the 1/2 MS medium containing NaCl. The differentiation status was investigated after 20–30 days.

2.3.7 Plantlets transplanting

Regenerated salt-tolerant plantlets were transplanted to normal soil (salt content lower than 0.3%) after adaptation at room temperature.

2.3.8 Haploid detection

The ploidy levels of the regenerated plants were determined by measuring the guard cell length of stomata on the leaves. The chlorophyll was first extracted completely with 70% ethanol. Then sections were prepared. The size of the stomata was measured using an eyepiece micrometer. The stomatal length of haploids ranged from 36 μm to 54 μm, which were 40%–50% shorter than that of diploids. The stomatal width of the haploids ranged from 18 μm to 24 μm. Based on the detection data, the regenerated plants were divided into haploids and naturally doubled diploids.

2.3.9 Chromosome doubling treatment on haploid seedlings

The most critical step in haploid breeding is the chromosome doubling. The following method was employed and over 70% chromosome doubled plants were obtained (Bai et al., 1979).

(1) Colchicine was dissolved in dimethyl sulfoxide (DMSO) to prepare a 0.4% colchicine solution.

(2) At active tillering stage, the roots of haploid seedlings were washed clean and the tillering nodes were immersed into the 0.4% colchicine solution. The treated seedlings were kept at 18℃ for 8 h to double the chromosomes. If the haploid seedlings had too many tillers, they were separated and put in tap water overnight before being treated in the colchicine solution so as to be immersed more thoroughly.

(3) The colchicine was rinsed out of the treated haploid seedling roots. Then the haploid seedlings were planted in common soil. The culture conditions were: night temperature 15℃, day temperature 18℃, 16/8 (light/dark) h photoperiod, and 70% humidity. Winter wheat needed a vernalization treatment (cultured at 4℃ for 28 days) and then seedlings were moved to normal temperature until the seed was set.

2.3.10 Identification of salt tolerance

The successfully doubled plants were grown for 3 continuous generations. Their agronomic characters were investigated. The F_4 generation was sown in the artificial salt pond (soil with 0.45% salt concentration) to identify salt tolerance. Those plants having stable salt tolerance were propagated. Physiological and biochemical detection methods were also used to explore the mechanisms of salt tolerance.

3. TECHNICAL EFFECTIVENESS AND APPLICATION STATUS

In vitro mutation is an economic and practical technique that combines *in vitro* culture and mutation induction. The use of the *in vitro* mutation technique in breeding enables to double and redouble the variation frequency, accelerate the purification speed, shorten the breeding period and reduce the population of the selection generation, thus improving the breeding efficiency.

This technique has wide application value and development prospects in crop breeding (Gao, 1992). Various mutation induction methods can be applied. For example, Hezu 8 is a high-yielding wheat variety, with early maturity, disease resistance, wet soil endurance and good quality, that was bred through the combination of nuclear mutation technology and *in vitro* culture (Gao, 1992). While *RH8706-49* a highly salt-tolerant wheat variety (salt tolerance index ⩾1.3) with many other characters of stress resistance was obtained by chemical mutation combined with *in vitro* culture (Shen et al., 1997). All this research has demonstrated that it is feasible to breed new salt-tolerant wheat varieties (or lines) via *in vitro* mutation technique.

In the protocols reported above, the following adjustments/modifications were included:

(i) the experimental procedure was simplified by adding a certain amount of mutagen to the induction medium to make the callus induction and mutation occur at the same time; (ii) the salt concentration chosen was higher than half lethal dose (LD_{50}) in order to achieve a high salt stress concentration so that the obtained salt-tolerant calli were from a genetically stable cell line, thus avoiding the result of physiological adaptation; and (iii) the selection effect was better if the NaCl stress was imposed at the callus differentiation stage rather than at the callus formation stage.

Moreover, experience demonstrates that further identification is needed to confirm whether the salt-tolerant mutants obtained via tissue and cell culture are real genetic mutations. Synthesizing the judgement criteria of Widholm, Flick and Maliga, one mutant can be identified as a true genetic mutant if: (i) this mutant occurs at low frequency; (ii) the variation characters are stable when away from the selection pressure; (iii) the callus induced from the regenerated plants with stable mutations should express the phenotype for which it was selected; (iv) the mutant can be sexually transmitted; and (v) there are changed gene products in the variant cells (Gao and Li, 2004). In our experiments, salt-tolerant variant plants were obtained from the differentiation medium containing 0.5% NaCl. After continuous planting in normal soil for two years, its progeny still had a higher level of proline as compared with the control. Especially under salt stress, the mutagenic progeny showed much better tolerance to salt, indicating that the obtained mutant had a certain stability of salt tolerance.

Since the advantages of artificial mutation combined with *in vitro* culture in salt-tolerant mutant selection were found, *in vitro* mutation technique has made good progress in breeding salt-tolerant mutants. Mutant plants have been obtained in many crops such as barley (Zhang et al., 1998), wheat (Shen et al., 1993), and triticale (Li et al., 1992). The mutagens involved are ^{60}Co γ rays, X rays, fast neutrons, Pingyangmycin (PYM), Daunomycin or Zhengdingmycin (ZDM), EMS, and sodium azide (NaN_3). In view of differential reaction of the explants to the mutagenic agents, explants can help determine which mutagen to use. For example, in these protocols, EMS was employed for mutation induction in anther calli whereas Pingyangmycin and Daunomycin were used to induce mutations in calli from mature embryos. The research also showed that it was easier to obtain salt-tolerant variations when using the genotypes that are more sensitive to the mutagens (Shen et al., 1997). Compound treatments can significantly increase the salt tolerance of the mutant.

In the process of selecting salt-tolerant mutants, *in vitro*-cultured organs and tissues are used to induce mutations. The occurrence of serious chimera makes it difficult to select and identify the mutants. The low mutation frequency greatly influences the selection efficiency too. Given the advantages and successful examples of the suspension cell culture in selecting salt-tolerant mutants, it is hoped that this technique will be used as the main technology to select salt-tolerant mutants in the future.

Finally, since the ultimate objective of breeding salt-tolerant wheat is to utilize the salt-affected lands and increase wheat yield in these lands, in the future salt tolerance should not be the only character to be studied, other agronomic characters should also be considered comprehensively in the selection of mutants. Therefore, the ideal salt-tolerant mutant must encompass a high comprehensive score in resistance, fertility, yield and quality. Along with the continuous development of modern science and technology, the combination of the somatic mutation breeding of salt tolerance with conventional breeding, MAS and genetic engineering can put together various fine traits to develop new varieties with enhanced salt tolerance, high-yield and good quality. The *in vitro* mutagenesis techniques will certainly be an effective breeding tool for the utilization of saline soils and will bring great economic benefits to the development of sustainable agriculture and forestry (Gao and Li, 2004).

REFERENCES

Bai, S. X., et al. (1979). A study of chromosome doubling of haploid plants of wheat (*Triticum vulgare*). *Acta Genet. Sin*, *6* (2), 230-232.

Bi, C. L., et al. (1999). Molecular Biological Identification of Wheat Salt-Tolerant Lines. *Hereditas*, *21* (6), 32-36.

Dai, G. X., et al. (2003). The Effects of Calcium on Salt tolerance in Plant. *Chin. Agric. Sci. Bull*, *19* (3) 97-101.

De, B. J., et al. (1987). 'Florin': A doubled haploid wheat variety developed by the anther culture method. *Plant Breed*, *98,* 53-56.

Gao, M. W., et al. (1992). a New Wheat Variety Developed with *in vitro* Mutation Technique. *Nucl. Phys. Rev*, *9*(4) 45-46.

Gao, Y. H., et al. (2004). Application of selection of salt-tolerant mutant from plant *in vitro*. *Acta Agric. Nucl. Sin, 18*(6), 448-452.

Hu, D. F., et al. (1983). The induction of pollen sporophyte of winter wheat and the development of the new variety "JINGHUA NO. 1". *Sci. Agric. Sin*, *1*, 29-35.

Li, S. P., et al. (1992). *In vitro* Culture, Callus Induction and Plant Regeneration of Hybrid Embryo Crossed between *Aegilops tauschii* and Hexaploid Triticolc. *Hereditas*, *2*, 137-144.

Melchers, G, (1972). Haploid higher plants for plant breeding. *Z. Pflanzenzuchtg*, *67,* 19-32.

Ouyang, J. W., et al. (1973). The induction of pollen regeneration plant and observing of the offspring in wheat. *Sci. Chin*, *1*, 72-78.

Shen, Y. Z., et al. (1993). A study of the salt-resistant variation of inducing matured embryo's callus and regenerated plants in wheat. *Acta Genet. Sin*, *20*(3), 283-291.

Shen, Y. Z., et al. (1997). A Study of salt-resistant variations induced in anther calli and regenerated plants in wheat. *Hereditas*, *6*, 7-11.

Shimada, T., et al. (1969). *In vitro* culture of wheat tissues: I. Callus formation, organ re-differentiation and single cell culture. *Can. J. Genet. Cytol*, *11* (3), 294-304.

Li, S. G., et al. (2001). Identification of the Molecular Markers Linked to the Salt-resistance Locus in the Wheat Using RAPD-BSA Technique. *Acta Bot. Sin*, *43* (6), 598-602.

Vinocur, B., & Altman, A. (2005). Recent advances in engineering plant tolerance to abiotic stress: achievements and limitations. *Curr. Opin. Biotechnol*, *16*, 1-10.

Wang, C. T., et al. (2001). Detection of the Wheat Salt-tolerant-mutant Using PCR-SSCP Combining with Direct Sequencing. *Acta Genet. Sin*, *22* (1), 10-13.

Wang, H. Y., et al. (2004). Mapping of Relative Salt Tolerance Gene in Wheat Salt-Tolerant Mutant Using Microsatellite Marker. *Acta Agron. Sin*, *30* (7), 679-699.

Zenk, M. H. (1974). Haploids in physiological and biochemical research, In Kasha K. J. (Eds.), *Haploids in higher plants: Advances and potential*, Guelph Univ., Guelph, Canada, 339-354.

Zhang, Y. L., et al. (1998). Selection and characterization of salt-resistant variants from *Hordeum brevisulatum*. *Pratacult. Sci*, *15* (1), 30-32.

Zhu, Z. Q., et al. (1973). Study on the induction of pollen regeneration plant and morphological forming procedure in wheat. *Acta Bot. Sin*, *15*, 1-11.

Appendix Ⅰ: Composition of culture media

1. FORMULAS OF MEDIA FOR MATURE EMBRYO CULTURE

(1) Induction medium: MS + 2,4-D 2.0 mg L^{-1} + 6-BA 0.5 mg L^{-1} + NAA 0.5 mg L^{-1} + LH 500 mg L^{-1} + sucrose 3%, pH 5.8. Mutagens Pingyangmycin (PYM) and Daunomycin or Zhengdingmycin (ZDM) were dissolved in distilled water and added to the medium as described.

(2) Subculture and selection medium: MS + 2,4-D 1.0 mg L^{-1} + LH 500 mg L^{-1} + yeast paste 1,000 mg L^{-1} + sucrose 3%, pH 5.8. NaCl at 0.5% and 1.2% was added to the medium for the selection of salt tolerance.

(3) Differentiation medium: 1/2 MS + 6-BA 1.0 mg L^{-1} + NAA 0.5 mg L^{-1} + sucrose 3%, pH 5.8. NaCl at 0.5%, 0.8% and 1.2% were added to the medium for a further selection of salt tolerance.

2. FORMULAS OF MEDIA FOR ANTHER CULTURE

(1) Induction medium: N6 + 2,4-D 2.0 mg L^{-1} + KT 0.5 mg L^{-1} + sucrose 9%.

(2) Subculture medium: N6 + 2,4-D 1.0 mg L^{-1} + KT 0.5 mg L^{-1} + sucrose 9%.

(3) Selection medium: N6 + 2,4-D 1.0 mg L^{-1} + KT 0.5 mg L^{-1} + NaCl 0.5% + sucrose 9%.

(4) Differentiation medium: N6 + KT 2.0 mg L^{-1} + IAA 0.3 mg L^{-1} + NaCl 0.5% + sucrose 3%.

Note: All the media were sterilized by autoclaving at 1.1 kg cm^{-2} for 20 min.

Abbreviations

2,4 D	2,4 Dichlorophenoxyacetic acid
6-BA	6-Benzylaminopurine
IAA	Indolacetic acid
KT	Kinetin
LH	Lactoprotein hydrolysate
NAA	Naphthalene acetic acid

Appendix Ⅱ: Pictures

Figure P1.1 Mature embryos of wheat on the induction medium supplemented with mutagens.

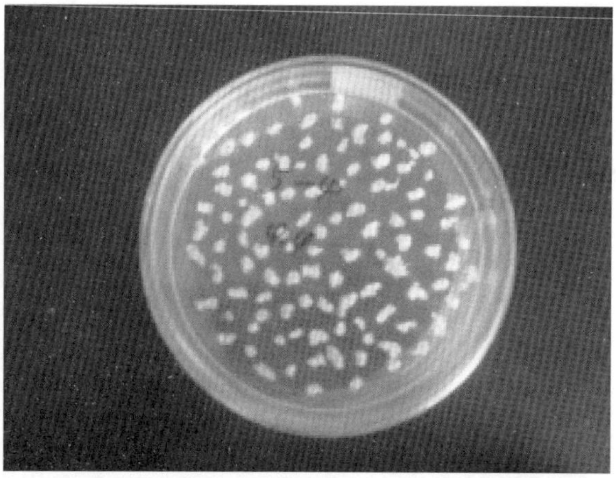

Figure P1.2 Calli induced from mature embryos of wheat.

Figure P1.3 Salt-tolerant calli on the differentiation medium containing salt.

Figure P1.4 Regenerated salt-tolerant plants growing in common soil.

Figure P1.5 Regenerated plants growing in the artificial salt pond at 0.45% salt concentration for identification of salt tolerance.

PROTOCOL Ⅱ

HAPLOID CELL ENGINEERING PROTOCOLS FOR BREEDING SALT-TOLERANT BARLEY

Jianhua Huang

Biotechnology Research Institute, Shanghai Academy of Agricultural Sciences, Shanghai 201106, China

1. INTRODUCTION

It is easier to obtain mutants for specific purposes from cell culture (Chaleff, 1983). Anthers contain a large number of haploid microspores, and the embryoids formed from the anther culture basically stem from the microspore de-differentiation, so the tolerance to NaCl of the barley anthers should be to a large extent related to the capacity of NaCl tolerance of the microspores. According to the scientific principles of cell totipotency and biological holography, the two requirements are as follows: (i) the survival of anthers (microspores predominantly) after NaCl stress should indicate that they contain the corresponding resistance gene; and (ii) the surviving haploid cells can be induced to diploid plants with the pure resistance gene by chromosome doubling.

The tolerance to NaCl of barley anthers (microspores) has been found consistent with the donor plants (Sun et al., 2006; 2007). This provides a theoretical basis for screening recombinant gametophytes and variants with NaCl resistance by using haploid cell engineering technology in barley. Selections of barley made at the haploid cell level can be propagated into double haploid stable salt-tolerant plants by chromosome doubling (Huang, 2007).

This contribution describes the protocols for screening barley doubled haploids with enhanced salt tolerance developed from barley anther culture employing radiation mutagenesis and NaCl treatment.

2. EXPERIMENTAL PROCEDURES

2.1 Anther culture and ^{60}Co irradiation mutagenesis

2.1.1 Growth of donor plants

The highest efficiency of barley anther culture is achieved when donor plants are grown in a controlled growth room at temperatures of about 18℃ during the day and 15℃ during the night, 16 h photoperiod (light intensity about 10,000 lux), and 60%–80% relative humidity. During the whole vegetation period, plants should be properly watered and fertilized to maintain vigorous growth.

If field-grown material is used as donor plants, the seeds should be sown well spaced in order to ensure that the plants have adequate space for growth and development. Addition of a small amount of nitrogen fertilizer at the meiosis stage will be beneficial to the anther culture.

2.1.2 Selection of donor spikes

In barley, the distance between the flag leaf and the penultimate leaf indicates the proper stage of microspore development. Generally speaking, tillers should be collected when the distance between the flag and the penultimate leaf is 2–3 cm. And the spikes in donor plants with anthers at the mid-uninucleate stage should be used for the cold pre-treatment.

2.1.3 Cold pre-treatment

Donor spikes wrapped with cling film are stored at 3℃ temperature for 15–20 days.

2.1.4 Spike irradiation mutagenesis and anther selection

After the cold pre-treatment, the spikes are treated by irradiation with ^{60}Co source at the dose rate of 1.2 Gy min^{-1} and a dose of 10 Gy.

Anthers containing microspores at the mid-uninucleate stage of the florets in the middle of spikes are selected for *in vitro* culture.

2.1.5 Induction medium

To the N6 medium used as a basic stock, add 2,4,5-T 2.0 mg L^{-1}, KT 0.2 mg L^{-1}, CH 800 mg L^{-1} and maltose 60 g L^{-1}, pH 5.8, solidified with 0.5% agar powder, load approximately 15 mL medium in 50 mL Erlenmeyer flasks separately, and sterilize them at 107.87 kPa for 15 min.

2.1.6 Anther culture

The tillers after the cold pre-treatment are put in a 75% ethanol solution for 3–5 min, then rinsed with sterile water 3–5 times. Anthers from the middle florets in each row of the spikes are used for culture *in vitro*. The anthers are removed from the florets in sterile conditions. The planting density should be 35–40 anthers per flask in 15 mL induction media, then the flasks are incubated at 20℃ in the darkness for 25–35 days.

2.1.7 Differentiation medium

To the C17 medium used as a basic stock, add 6-BA 0.5 mg L^{-1}, KT 1.5 mg L^{-1}, NAA 0.1 mg L^{-1} and maltose 30 g L^{-1}, pH 5.8, solidified with 0.5% agar powder, load approximately 25 mL medium in the 100 mL Erlenmeyer flasks separately, and sterilize them at 107.87 kPa for 15 min.

2.1.8 Differentiation culture of callus

Calli of 1–2 mm in diameter are transferred into the Erlenmeyer flasks containing 25 mL differentiation medium, at a density of 8–10 structures per flask. The Erlenmeyer flasks with calli are incubated at 25℃ and 11 h photoperiod.

2.1.9 Healthy culture medium

To the 1/2 MS medium used as a basic stock, add chlorocholine chloride 5.0 mg L^{-1} and sucrose 20 g L^{-1}, pH 6.0, solidified with 0.6% agar powder, load approximately 30 mL medium in the 100 mL Erlenmeyer flasks separately, and sterilize them at 107.87 kPa for 15 min.

2.1.10 Seedlings growth in healthy culture

Green seedlings are chosen for transfer onto the healthy culture medium, at a density of 3–4 seedlings per bottle, at 20℃ and 11 h photoperiod for 20–25 days. Then the bottles with the seedlings are incubated at 5℃ in natural light.

2.1.11 Growth of plants in soil

When roots have developed, the plants are transferred to pots containing garden soil and grown at 20℃/15℃ (day/night) and 11 h photoperiod. During the prophase 2–3 sprays should be used to keep wet the plants. Ensure that adequate nitrogen, phosphorus and potassium fertilizers are added at both tillering and jointing stages.

Because about 80% of the anther culture derived-barley plants obtained with this protocol are spontaneously doubled and completely fertile, a colchicine treatment is not required.

2.2 Barley seed mutagenesis and field selection

Dry barley seeds are treated by irradiation with a ^{60}Co source at the dose of 200 Gy. The treated seeds are then sown in the field, harvesting one spike per plant at harvest time. The seeds are sown, recording each spike, in a cool climate environment such as the mountainous areas of Yunnan. At maturity, choose the plants with excellent variations, harvest the seeds keeping records of each plant, and then sow them in the field separately.

2.3 Barley anther culture–NaCl stress screening procedures

2.3.1 Donor plants

Use excellent combinations of F_1 hybrid plants, and the fine mutations derived from mutagenesis.

2.3.2 Anther culture–NaCl stress medium

Similar procedure as described in 2.1 plus 0.3% NaCl.

2.3.3 Differentiation–NaCl stress medium

Similar procedure as described in 2.1 plus 0.5% NaCl.

2.3.4 Anther culture and callus differentiation under NaCl stress

Similar procedure as described in 2.1.

2.4 Identification of salt tolerance in barley seedlings at the germination stage

2.4.1 Seed germination medium

1.5% NaCl solution, pH 6.0, solidified with 6% agar powder, load approximately 30 mL medium in 100 mL Erlenmeyer flasks separately.

2.4.2 Seed pre-treatment

Seeds are soaked in 0.5% NaCl solution for 24 h.

2.4.3 NaCl stress culture

Germinated seeds are put onto NaCl stress medium, 5 seeds per bottle, 5 replicates, at 28℃ for 5 days.

2.4.4 Selected traits record

Radicle (R) length, coleoptile (C) length and the number (N) of roots are measured in control and salt stress treatments, separately.

2.4.5 *Standard salt tolerance evaluation*

Using the data recorded, the evaluation is made as follows:

$$\text{Radicle growth rate} = [(RS-RC)/RC] \times 100 \qquad (P2.1)$$

where *RS* is the average length of radicle in salt stress; *RC* is the average length of radicle in control.

$$\text{Coleoptile growth rate} = [(CS-CC)/CC] \times 100 \qquad (P2.2)$$

where *CS* is the average length of coleoptile in salt stress; *CC* is the average length of coleoptile in control.

$$\text{The number of roots growth} = [(NS-NC)/NC] \times 100 \qquad (P2.3)$$

where *NS* is the average number of roots in salt stress, *NC* is the average number of roots in control.

2.5 Heredity improvement methodology of barley double haploids for salt tolerance

Figure P2.1 shows the procedures followed for obtaining barley double haploids with improved salt tolerance.

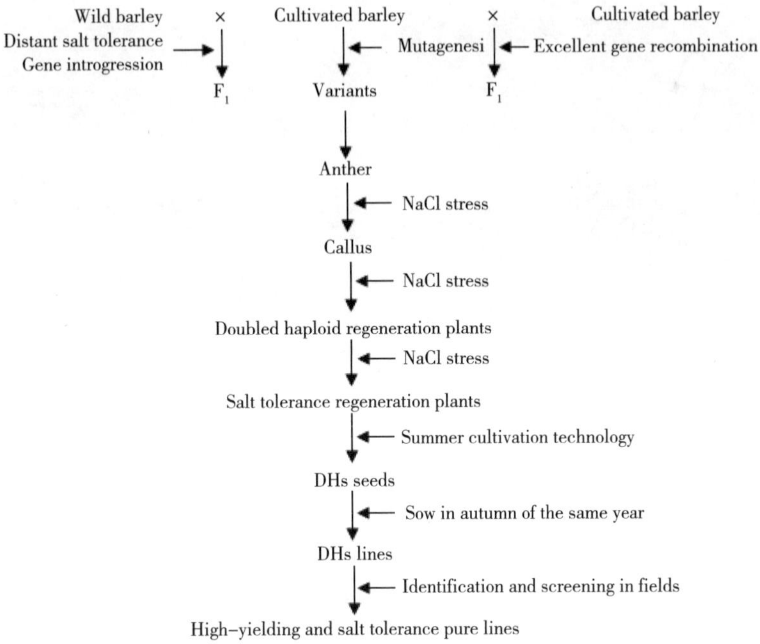

Figure P2.1 Heredity improvement technology procedures of barley double haploids for salt tolerance.

3. APPLICATION AND EFFICIENCY OF THE TECHNOLOGY

Similar to a population of the F_2 generation grown in the field, the population of the F_1 generation derived from a microspore culture is also one of the populations with the most variable genotypes. The genotypes with relevant resistance genes can be screened out under NaCl stress through the *in vitro* anther culture system (Figure P2.2, Figure P2.3). The embryoids derived from these microspores are then induced to differentiate seedlings under stress again (Figure P2.2). The basic composition of the culture media is shown in Table P2.1. The seedlings with relevant resistance genes are induced to doubled haploids (DH) plants, and immediately become homozygous lines. Excellent plants can be derived from the screening of DHs under salt stress. These plants are cultivated in the summer of the same year, and the seeds are harvested as pure lines in autumn. These pure breeding lines can be screened out through field identification in the spring of the second year. Comparative varietal testing can take place in the following autumn, and then the official field test can be arranged in the third year. Homozygous breeding lines with excellent genes originating from four different parents separately can be derived in two years through this breeding system. These fine DHs varieties always possess the characters of high purity. The chromosome configuration of the doubled haploids is shown to be quite stable through karyotype analysis.

Figure P2.2 Regenerated plants.

审定证书

（认定）

上海市农科院生物技术中心

浙江省嘉兴市农科院

你单位培育(引进)的 花22 品种，已经第四届上海市农作物品种审定委员会第九次会议审定通过，特发此证。

品种来源：自选 秀麦1号/秀麦3号 F2代离体培养

审定编号：沪农品认麦（2006）第002号

二00六年二月二十三日

Figure P2.3 Certificate of variety Hua-22.

The two most important characters of the doubling haploids in the heredity improvement include: i) we only need to select parents with genes of the target traits, irrespective whether the parents are homozygous or not; and ii) we can make the efficiency selection in DH_1 generation.

Table P2.1 Basic composition of three common plant tissue and cell culture media

Mineral salts	MS		Mineral salts	N6	Mineral salts	C17
Macroelements	mg L^{-1}	mmol L^{-1}	Macroelements	mg L^{-1}	Macroelements	mg L^{-1}
NH_4NO_3	1650	20.6	KNO_3	2830	NH_4NO_3	200
KNO_3	1900	18.8	$(NH_4)_2SO_4$	463	KNO_3	300
$CaCl_2 \cdot 2H_2O$	440	3.0	$CaCl_2 \cdot 2H_2O$	166	KCl	150
$MgSO_4 \cdot 7H_2O$	370	1.5	$MgSO_4 \cdot 7H_2O$	185	$CaCl_2 \cdot 2H_2O$	250
KH_2PO_4	170	1.25	KH_2PO_4	400	$MgSO_4 \cdot 7H_2O$	325
					KH_2PO_4	150
Microelements	mg L^{-1}	μmol L^{-1}	**Microelements**	mg L^{-1}	$NaH_2PO_4 \cdot H_2O$	100
KI	0.83	5.0	$FeSO_4 \cdot 7H_2O$	27.8		
H_3BO_3	6.2	100	Na_2EDTA	37.3	**Microelements**	mg L^{-1}
$MnSO_4 \cdot 4H_2O$	22.3	100	$MnSO_4 \cdot 4H_2O$	4.4	Na-Fe-EDTA	17.5
$ZnSO_4 \cdot 7H_2O$	8.6	30	$ZnSO_4 \cdot 7H_2O$	1.5	Fe-citrate	3.0
$Na_2MoO_4 \cdot 2H_2O$	0.25	1.0	$Na_2MoO_4 \cdot 2H_2O$	0.25	$MnSO_4 \cdot 4H_2O$	0.5
$CuSO_4 \cdot 5H_2O$	0.025	0.1	KI	0.8	$ZnSO_4 \cdot 7H_2O$	0.25
$CoCl_2 \cdot 6H_2O$	0.025	0.1	H_3BO_3	1.6	$CoCl_2 \cdot 6H_2O$	0.012
Na_2EDTA	37.3	100			$CuSO_4 \cdot 5H_2O$	0.012
$FeSO_4 \cdot 7H_2O$	27.8	100			$Na_2MoO_4 \cdot 2H_2O$	0.012
					KI	0.1
					H_3BO_3	5.0
Sucrose(g)	30					
pH	5.7					

Through the application of the above procedures, a group of barley germplasm materials with markedly improved salt tolerance has been selected. A number of salt-tolerant high-yielding new lines have been bred through field identification and comprehensive selection for agronomic traits. For instance, Hua-22 is one of these new barley varieties (Figure P2.3) showing high-yield and quality, large ears and middle-stalked, 85 cm of plant height, 7.2 cm of ear length, 24 grains per panicle, about 48 g of 1,000-grain weight, lenient and green leaves, strong middle-tillers, strong stem, vigorous growth, a high rate of spike and seed-setting, kernels and aristae easily falling at harvest. It is suitable for cultivation in coastal areas and has been cultivated in an area of nearly 0.33 Mha in Shanghai and Jiangsu province (Figure P2.4). This variety belongs to the two-row long awn hull kernel type spring barley, with strong salt tolerance and restoration capability after freezing damage, strong lodging resistance, low protein content (about 10.99%), and good quality for beer, moderate resistance to scab and powdery mildew, yellow mosaic disease tolerance, but slight susceptibility to barley stripe disease.

Figure P2.4 Field-growing of barley Hua-22.

REFERENCES

Chaleff, R. S., et al. (1983). Isolation of agronomically useful mutants from plant cell cultures. *Science*, *219*, 676-682.

Huang, J. H., et al. (2007). Application of microspore and anther culture to directional genetic improvement in barley. *Crop Res*, *3*, 167-169.

Sun, Y. F., et al. (2006). Effects of NaCl stress on anther culture for different genotypes in barley. *J. Nucl. Agric. Sci*, *20*(1), 19-22.

Sun, Y. F., et al. (2007). Response of *in vitro* Culture of Anther and Microspore from Different Genotypes of Barley to Cool and NaCl pre-treatments. *Chin. Agric. Sci. Bull*, *23*(4), 46-48.

PROTOCOL Ⅲ

IN VITRO MUTAGENESIS PROTOCOLS FOR THE SELECTION OF SALT-TOLERANT RICE

Don Sub Kim, Jae Young Song, Jin-Baek Kim, Geung-Joo Lee and Si-Yong Kang[*]

Advanced Radiation Technology Institute, Korea Atomic Energy Research Institute (KAERI)

[*]Correspondence: sykang@kaeri.re.kr

1. INTRODUCTION

Radiation-induced mutagenesis has been extensively studied in the identification and isolation of plant genes and in plant breeding programs. Rapid progress in plant tissue culture methods is attributed to the large availability of mutants that can be generated for specific biochemical events making it difficult or impossible to use them as intact plants (Carlson, 1970; Larkin and Scowcroft, 1981; Schaeffer and Sharpe, 1983; Kim et al., 2004). However, this limitation can be overcome by co-culturing plant tissues with pathogens or their toxins and/or manipulating a medium composition, e.g. selection for resistance to salinity, drought and amino acid analogs.

Although it is possible to select mutants with salt tolerance by somaclonal variation, the mutant frequencies in the plant systems are often low due to inefficient recovery techniques, the presence of multiple gene copies located on different chromosomes, and the occurrence of cell-to-cell interactions even the most ideal suspension culture systems (Schaffer and Sharpe, 1983). Therefore, *in vitro* mutation induction techniques provide tools for rapid mutant creation as well as an increase in the selection frequencies of cell types, for which there is no obvious counterpart in field environments (Maluszynski et al., 1995). Such techniques enable a greater use of mutated genes for crop improvement and their cumulative effects may result in more desirable characters than those of the original variety.

Rice is an important staple food, which is consumed by more than half of the world's population. However, rice is a crop sensitive to salt stress. Therefore, the development of rice varieties with increased salt tolerance is needed to increase grain yields and expand its cultivation in salt-affected lands. The possibility of screening for salt stress at the tissue culture level makes *in vitro* methods attractive for developing salt-tolerant rice plants (Lee et al., 2003).

This contribution describes the methodologies followed in the selection of radiation-induced salt-tolerant rice mutants through *in vitro* mutagenesis. A brief assessment of the efficiency and application of the technique is also included.

2. EXPERIMENTAL PROCEDURES

The present protocols include the following main steps: (i) induction and selection of salt-tolerant cell lines via *in vitro* mutagenesis; (ii) plant regeneration and a subsequent selection through each generation (M_2-M_3); and (iii) screening of salt-tolerant rice lines in a saline field. Figure P3.1 shows a flow chart of the general procedures followed in the selection of salt-tolerant rice lines.

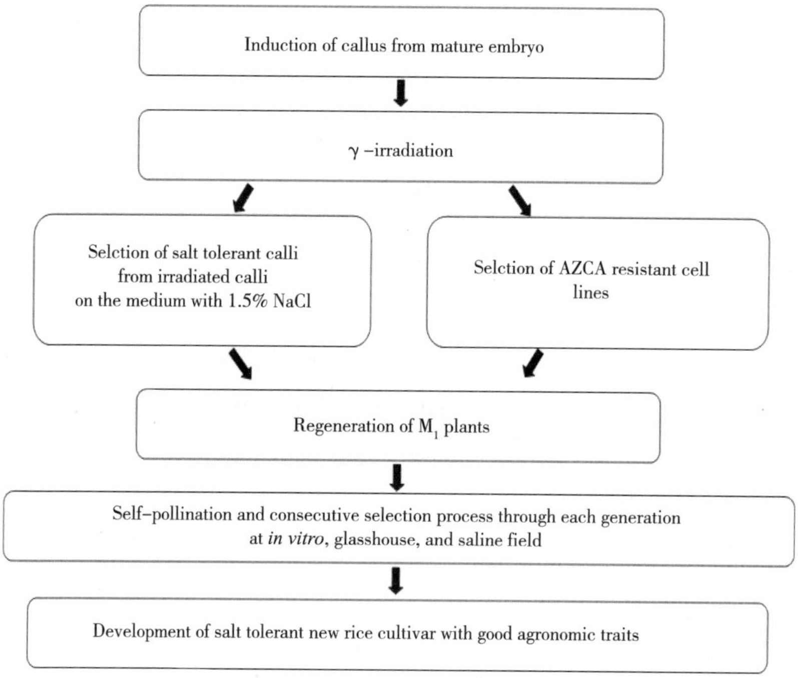

Figure P3.1 Flow chart of procedures for the selection of salt-tolerant mutant rice lines.

2.1 *In vitro* selection of salt-tolerant rice mutant cell lines by radiation mutagenesis

2.1.1 Embryogenic callus induction

Calli were initiated from embryos of hulled seeds of a japonica rice cultivar Dongjinbyeo is a high-yielding and quality cultivar with lodging and disease resistance. Seeds were sterilized in 10% sodium hypochlorite with 2–3 drops of Tween-40 for 30 min. The embryos were placed on a callus induction medium (N6 basal medium supplemented with 2 mg L^{-1} 2,4-D, 30 g L^{-1} sucrose and 0.4% (W/V) phytagel) in sterile petri dishes, and incubated in the dark at (25±1)℃ for callus initiation. The medium (Table P3.1) was adjusted to pH 5.8 and autoclaved for 20 min at 1.1 kg cm^{-2} pressure at 121℃. Induced calli were once more subcultured for proliferation with 4 weeks intervals.

Table P3.1 Composition of callus induction and regeneration media for *in vitro* selection of salt-tolerant cell lines

Medium type	Components
Callus induction	N_6 basal, 30 g L^{-1} sucrose, 2 mg L^{-1} 2,4-dichlorophenoxyacetic acid (2,4 D), pH 5.7–5.8, 4 g L^{-1} phytagel
Callus selection	Similar to the callus induction medium but with 1.5% NaCl
Plant regeneration	MS basal, 30 g L^{-1} sucrose, 0.5 mg L^{-1} Naphthaleneacetic acid (NAA), 2 mg L^{-1} Benzylaminopurine (BAP), pH 5.7–5.8, 4 g L^{-1} phytagel
Plant selection	MS basal, 30 g L^{-1} sucrose, 1.0% NaCl, pH 5.7–5.8, 4 g L^{-1} phytagel
Root formation	Similar to the plant selection medium, but no NaCl

2.1.2 Determining optimum NaCl concentration and irradiation dose for salt-tolerant calli

It is very important to decide on the optimal salt concentration and the irradiation dose for the selection of salt-tolerant calli. The calli were divided into small pieces (0.5–1 mm diameter) and inoculated on a N6 medium containing increasing NaCl concentrations (0, 0.25%, 0.5%, 0.75%, 1.0%, 1.25% and 1.5%) and the survival rate and fresh weight of the surviving calli for 40 days were investigated. The callus pieces were irradiated with 0 Gy, 30 Gy, 50 Gy, 70 Gy, 90 Gy, and 120 Gy γ-rays from a ^{60}Co source, and the fresh weight and survival rate of calli for 40 days were also investigated (Figure P3.2A). In each experiment, there were three replicates for each treatment.

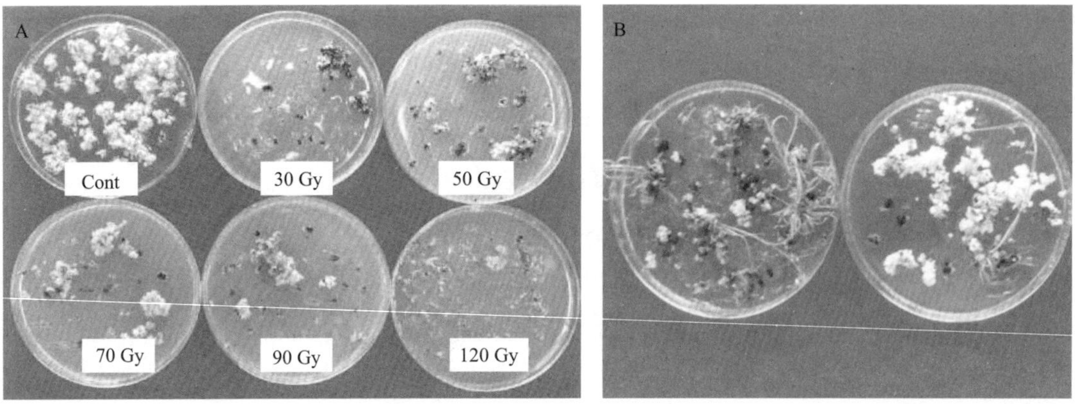

Figure P3.2 (A) *Growth comparison of calli irradiated with various radiation doses growing on the medium with 1.5% NaCl. Control plate containing non-mutagenized calli growing in the medium without salt.* (B) *Salt-tolerant calli growing in the regeneration medium.*

2.1.3 Selection for NaCl tolerance

The propagated calli were divided into small pieces (0.5–1 mm diameter) and inoculated on a N6 medium treated with 1.5% NaCl and 70 Gy irradiation, which were found the optimum NaCl concentration and irradiation dose for the selection of salt-tolerant calli, respectively (Table P3.1). The selected salt-tolerant calli were maintained at the same concentration.

2.1.4 Plant regeneration

The selected NaCl tolerant calli were transferred and cultured for regeneration on a MS medium supplemented with 0.5 mg L^{-1} NAA and 2 mg L^{-1} BAP, but without salt for 30 days (Table P3.1, Figure P3.2B). Plantlets were cultured at 10 cm on a half-MS medium followed by an acclimation at room temperature. Some of the regenerated plants were transferred to a medium with 0.75 % NaCl to confirm a transmission of the salt tolerance from the callus to the regenerated plants. Numbers were assigned to each of the regenerated plants (M_1). Plants were grown to maturity in a field having NaCl-free soil. Standard crop management practices, which included a fertilizer application(N-P-K) of 11-7-8 kg /10 a were followed. Harvested grain of the lines was sterilized with a seed disinfectant.

2.2 *In vitro* selection of AZCA-resistant mutant cell lines

When plant cells are exposed to stresses such as drought, salinity and cold, a proline accumulation is a widespread phenomenon. Under stress conditions a proline increase plays a key role for osmotic adjustments in a large number of plant species (Delauney and Verma, 1993). Mutants resistant to proline analogs are thought to be useful material for studying proline biosynthesis and saline, drought, and cold resistance in higher plants (Miflin et al., 1983).

The proline analog Azetidine-2-carboxylic acid (AZCA) is a natural product found in some species of *Liliaceae*, which inhibited irreversibly cell growth by competition with the proline for its incorporation into the proteins (Cella et al., 1982; Hyun et al., 2003). This incorporation presumably leads to an altered protein conformation and function, and accounts for the cyto-toxic effects of AZCA (Lodato et al., 1984; Song et al., 2007). In many cases, the resistance mechanism to amino acid analogs has been shown to be due to the insensitivity to feedback inhibition (Widholm, 1974). Mutants with an altered feedback mechanism express more resistance to feedback inhibition resulting in a greatly elevated free proline accumulation. The decrease of the uptake inhibitors by a change in cell membrane permeability and an inactivation of inhibitors have also been reported as mechanisms of resistance to an amino acid analog.

2.2.1 Embryogenic callus induction

Dehusked seeds of the cultivar Donganbyeo were sterilized in 70% ethanol for 30 s followed by an addition of 5% sodium hypochlorite with 2–3 drops of Tween-20 for 20 min. The embryos of seeds were rinsed 3–4 times with sterile distilled water and placed on a N6 medium supplemented with 2 mg L^{-1} 2,4-D, 30 g L^{-1} sucrose and 0.4% (W/V) phytagel in sterile petri dishes, and incubated in the dark at (25±1)℃ for callus initiation. The medium was adjusted to pH 5.8 and autoclaved for 20 min at 1.1 kg cm^{-2} pressure and 121℃. Induced calli were subcultured once more for proliferation.

2.2.2 Determining the optimum AZCA concentration

Induced callus was divided into 30 pieces (about 1 – 2 mm) with a sterile razor and transferred to the N6 basal medium containing 1.5 mmol L^{-1}, 2 mmol L^{-1}, 3 mmol L^{-1}, and 4 mmol L^{-1} AZCA concentrations. Thirty callus pieces each were plated per concentration with 10 replications. The data on the survival rate and fresh weight was obtained 40 days after culture. Based on the survival rate and fresh weight, the optimum AZCA concentration was assessed as 3 mmol L^{-1}or 4 mmol L^{-1}.

2.2.3 γ-ray irradiation treatment and selection of AZCA-resistant cell lines

The effect of γ-rays on growth of the callus was investigated 40 days after an irradiation treatment. Calluses were irradiated with 0 Gy, 30 Gy, 50 Gy, 70 Gy, 90 Gy, and 120 Gy γ-rays. Selection of AZCA-resistant cell lines was conducted on a selection medium containing 3 mmol L^{-1}or 4 mmol L^{-1} AZCA.

2.2.4 Regeneration of AZCA-resistant cell lines

AZCA-resistant calluses were transferred to the MS medium containing 0.2 mg L^{-1} indolacetic acid (IAA), 3 mg L^{-1} kinetin, and 30 g L^{-1} sucrose solidified with 0.5% (W/

V) phytagel. The medium was adjusted to pH 5.8 and autoclaved for 20 min at 1.1 kg cm^{-2} pressure and 121℃. Regenerated plants were obtained at 30 days after being transferred to the regeneration medium. The plantlets were grown further in a bottle containing a half strength MS medium without hormone. Regenerated plantlets from the resistant callus were grown in a solution containing a half strength MS basal and 2 mmol L^{-1} AZCA to determine their resistance. When the seedlings were at the 2–3 leaf stage, the tops of the shoots were removed except for 5 cm in length and the roots were removed thoroughly for a comparison of their regrowth ability. AZCA-resistant regenerants were selected after 2 weeks.

2.3 Subsequent selection of salt-tolerant rice lines at *in vitro*, glasshouse and field

2.3.1 Selection of the optimum salt concentration at a plant level

It is very important to decide on the concentration resulting in 50% mortality (LD_{50}) for each cloned plant under a salt concentration. Hulled seeds were sterilized with 5% sodium hypochlorite, and cultured on a half strength MS agar medium containing 0%, 0.25%, 0.5%, 0.75%, 1.0%, 1.5% and 2.5% NaCl concentrations for 40 days. During this period, the highest efficiency of plant culture was achieved when the plants were grown in a controlled growth room at (27±1) ℃ for 16 h / 8 h (light / dark), for the determined optimum salt concentration

2.3.2 M_2 generation and screening of the M_3 seedlings

Three hundred M_2 lines from M_1 plants, except those with a poor plant type (droopy leaves and weak culm), were harvested and numbered from M_2-1 to M_2-300. The M_2 generation was grown as a plant in rows in a NaCl-free field and harvested as M_3 generation (5,000 lines) showing normal grain fertility (above 80%) on an individual plant basis. Each M_3 line was the progeny of a single M_2 line seed bulk and numbered from M_3-1-1 to M_3-1-*n* and from M_3-300-1 to M_3-300-*n* (*n* was influenced by lines).

M_3 lines were screened for salinity tolerance *in vitro* and in a glasshouse with trays in nutrient solution with 0.75% NaCl (Figure P3.3A, Figure P3.3B). In the glasshouse, each tray had 20 pots (60 mm × 150 mm × 30 mm), with one from the parent, not cultured *in vitro*. The trays were filled with fine soil, which was commercially made for rice culture, and 80 seeds per line were placed in each pot at a depth of 5 mm. The trays were watered with tap water until the 3–4 leaf stage. At that stage, excess water was drained, and the trays with rice seedlings were refilled with the nutrient solution containing 0.75% (EC_e = 13 dS m^{-1}) salt (NaCl) and 1 g L^{-1} fertilizer. The solution was well mixed with a stirrer to maintain a uniform salt concentration. The EC_e measurements were taken daily. After three weeks growth under salt stress (EC_e = 13 dS m^{-1}), salinity symptoms were scored according to the Standard Evaluation System (1–3: tolerant, 5: moderate, 7–9: sensitive) developed at IRRI. To re-estimate the salt tolerance, thirty seeds each of the tolerant and sensitive lines were placed in glass bottles (5 cm × 7 cm) with a salt-free solution and 0.75% salt treatment and cultured for 30 days and 3 weeks, respectively. Plant height was used to assess the salt tolerance or sensitivity of the rice lines for quantitative measurement of their salt tolerance.

Figure P3.3 Screening of selected salt-tolerant rice mutant lines in vitro (A), *glasshouse* (B), *and saline field* (C).

2.3.3 Selection of salt-tolerant lines in a saline field

M_3 or M_4 lines selected at the seedlings stage were tested in a saline field near the sea coast (Figure P3.3C). The salinity level of the saline field was measured (EC_e=10–14 dS m^{-1}). The heterogeneity of the soil salinity in the field was an inherent problem for evaluating the performance of the lines. However, this problem was minimized with plot sizes of 2.4 m × 4.5 m. From each line, 30 plants were transplanted in two replicates at 30 cm × 15 cm apart in both saline and normal fields. The original variety was used as the control, and transplanted once every 8 lines. Standard crop management practices including a fertilizer application (N-P-K) of 20-8-8 kg /10 a were followed. Grain was harvested on an individual plant basis.

2.3.4 Performance evaluation at maturity

All lines were tested in a field trial under saline and non-saline conditions in five replications, and evaluated at maturity (Figure P3.4). Plant height, panicle length, tiller number, survival rate, number of spikelets per plant and grain weight were recorded. Analysis of the variance and means comparisons were made with the data from the saline and non-saline fields. The *t*-test value at 5% and 1% level of significance was used to determine the superiority of the tested lines over the control.

Figure P3.4 Comparison of agronomic characteristics of M_3 lines cultured in the 0.8% saline (A) *and normal (non-saline) fields at the Gyehwado Substation of the Honam Agricultural Research Institute, Rural Development Administration, Rep. of Korea.*

3. EFFICIENCY AND APPLICATIONS

Variations in the progenies of the *in vitro* mutagenized plant materials have already been reported. They probably have originated from the genetic instability created during plant tissue culture in combination with induced mutations (Maluszynski et al., 1995).

In vitro culture in combination with γ-ray induced mutations is an effective way to improve the salt tolerance in rice. The mutation and selection frequency were greatly increased, and the selected mutants were stable. Such mutant lines could be further utilized as a source of salt-tolerant germplasm. A new rice cultivar Wonhaebyeo with salt tolerance was developed by KAERI in 2007 using the above mentioned *in vitro* mutagenesis with 70 Gy γ-rays. This rice mutant will be released for cultivation in the reclaimed salt-affected lands and also used as a control plot for genetic research on salt tolerance.

REFERENCES

Carlson, P. S., et al. (1970). Induction and isolation of auxotrophic mutants in somatic cell cultures of *Nicotiana tabacum*. *Science*, *168*, 487-489.

Cella, R., et al. (1982). Characterization of a carrot cell line resistant to azetidine-2-carboxylic acid. *Plant Sci. Lett*, *24*, 125-135.

Delauney, A. J., et al. (1993). Proline biosynthesis and osmoregulation in plants. *Plant J*, *4*, 215-223.

Hyun, D. Y., et al. (2003). Selection of azetidine-2-carboxylic acid resistant cell lines by *in vitro* mutagenesis in rice (*Oryza sativa* L.). *J. Plant Biotech*, *5*(1), 43-49.

Kim, D. S., et al. (2004). Selection of radiation induced 5-methyltryptophan resistant rice mutants through embryo culture. *Euphytica*, *135*(1), 9-19.

Larkin, P. J., et al. (1981). Somaclonal variation–a novel source of variability from cell cultures for plant improvement. *Theor. Appl. Genet*, *60*, 197-214.

Lee, I. S., et al. (2003). Selection and characterizations of radiation-induced salinity tolerant lines via rice embryo culture. *Breed Sci*, *53*(4), 313-318.

Lodato, R. F., et al. (1984). Mutant cell lines resistant to azetidine-2-carboxylic acid: alterations in the synthesis of proline from glutamic acid. *J. Cell Physiol*, *119*, 137-143.

Maluszynski, M., et al. (1995). Application of *in vitro* mutation techniques for crop improvement. *Euphytica*, *85*, 303-315.

Miflin, B. J., et al. (1983). Amino acids, nutrition and stress: the role of biochemical mutants in solving problems of crop quality, In Kosuge, T., Meredith, C.P., & Hollaender, A. (Eds.), *Genetic engineering of plants - an agricultural perspective*, Plenum Press, London, 391-414.

Schaeffer, G. W., et al. (1983). Mutations and cell selections: genetic variation for improved protein in rice, In Owens, L. (Eds.), *Genetic Engineering: Applications to Agriculture*, Proc. Symp., Rowman & Allanheld, Totowa, NJ 237-254.

Song, J. Y., et al. (2007). Characterization of salt-tolerant rice mutant lines derived from Azetidine-2-carboxylic acid resistant cell lines induced by gamma ray irradiation. *J. Plant Biotech*, *34*(1), 61-68.

Widholm, J. M., et al. (1974). Cultured carrot cell mutants: 5-methyltryptophan-resistant trait carried from cell to plant and back. *Plant Sci. Lett*, *3*, 323-330.

APPENDIX

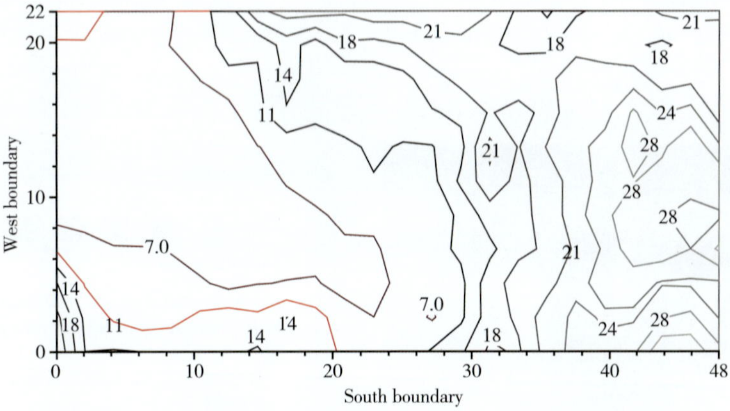

Note: ①Former name, now renamed as Khyber Pakhtunkhwa Province.

Figure 2.10 Example of the type of plot produced of salinity levels in the field [Gundheri, NWFP (Northwest Frontier Province)① *Pakistan, 13 November 1997]. The plotted data are* EC_e (ds m^{-1})*, produced from the* EC_a *after calibration of the EM38.*

Figure P1.1 Mature embryos of wheat on the induction medium supplemented with mutagens.

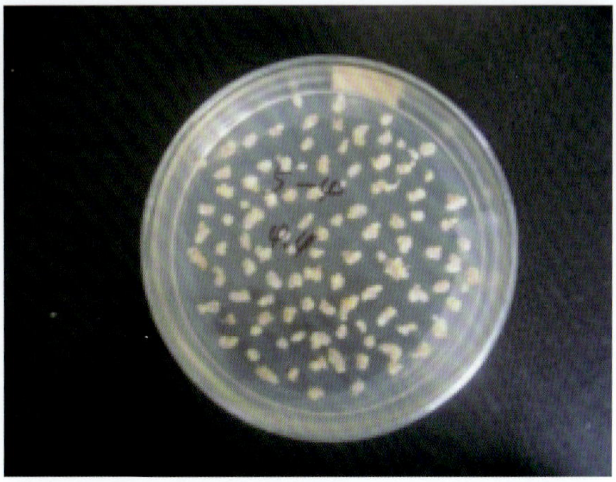

Figure P1.2 Calli induced from mature embryos of wheat.

Figure P1.3 Salt-tolerant calli on the differentiation medium containing salt.

Figure P1.4 Regenerated salt-tolerant plants growing in common soil.

Figure P1.5 Regenerated plants growing in the artificial salt pond at 0.45% salt concentration for identification of salt tolerance.

Figure P2.2 Regenerated plants.

审 定 证 书

（认 定）

上海市农科院生物技术中心
浙江省嘉兴市农科院

你单位培育(引进)的 花22 品种，已经第四届上海市农作物品种审定委员会第九次会议审定通过，特发此证。

品种来源：白荡 秀麦1号/秀麦3号 F2代离体培养

审定编号：沪农品认麦（2006）第002号

二00六 年 二 月

Figure P2.3 Certificate of variety Hua-22.

Figure P2.4 Field-growing of barley Hua-22.

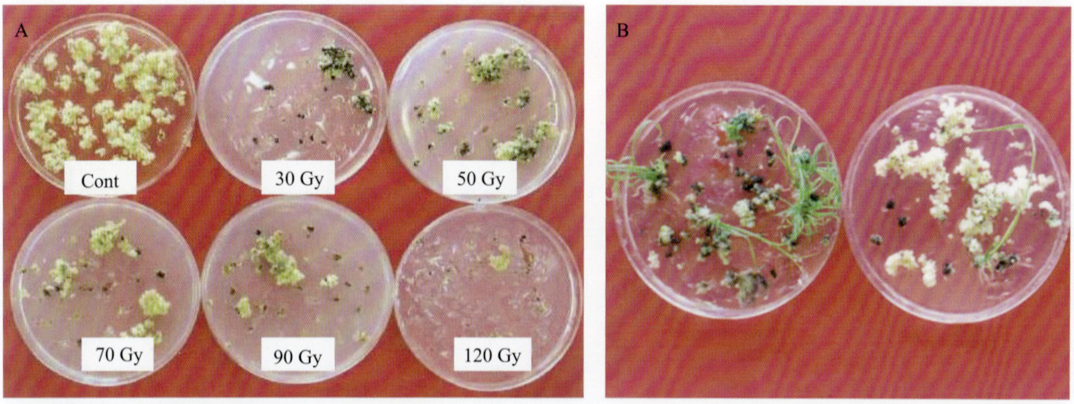

Figure P3.2 (A) *Growth comparison of calli irradiated with various radiation doses growing on the medium with 1.5% NaCl. Control plate containing non-mutagenized calli growing in the medium without salt.* (B) *Salt-tolerant calli growing in the regeneration medium.*

Figure P3.3 Screening of selected salt-tolerant rice mutant lines in vitro (A), *glasshouse* (B), *and saline field* (C).

Figure P3.4 Comparison of agronomic characteristics of M_3 lines cultured in the 0.8% saline (A) *and normal (non-saline) fields at the Gyehwado Substation of the Honam Agricultural Research Institute, Rural Development Administration, Rep. of Korea.*